T0390075

Active Site-directed Enzyme Inhibitors

Design Concepts

Active Site-directed Enzyme Inhibitors
Design Concepts

By

Weiping Zheng
Jiangsu University, China
Email: wzheng@ujs.edu.cn

Print ISBN: 978-1-83916-197-1
PDF ISBN: 978-1-83916-766-9
EPUB ISBN: 978-1-83767-143-4

A catalogue record for this book is available from the British Library

The Royal Society of Chemistry is a charity, registered in England and Wales, Number 207890, and a company incorporated in England by Royal Charter (Registered No. RC000524), registered office: Burlington House, Piccadilly, London W1J 0BA, UK, Telephone: +44 (0)20 7437 8656.

Visit our website at www.rsc.org/books

Printed in the United Kingdom by CPI Group (UK) Ltd, Croydon, CR0 4YY, UK

Preface

There have been many concepts (catalytic mechanism-based and substrate-based designs, *etc.*) whose exploitation has been demonstrated to be able to quickly and cost-effectively afford effective active site-directed inhibitors for various enzymatic reactions involved in various life processes and therapeutic areas. This book will be the first of its kind in the field, introducing to its audience in a single book fashion these concepts succinctly yet comprehensively. The book is organized by concept and for each concept there is a delineation of its mode of working and its applications with different types of enzymatic reactions. I hope this book will benefit its audience for quickly and efficiently obtaining effective active site-directed inhibitors for any of the enzymatic reactions under study without a need to resort to library screening- and biostructure-based techniques. In addition to its use as a reference book, this book could also be adopted as a text in a course for graduate students or upper-level undergraduate students.

Weiping Zheng

Active Site-directed Enzyme Inhibitors: Design Concepts
By Weiping Zheng

Published by the Royal Society of Chemistry, www.rsc.org

Contents

Active Site-directed Enzyme Inhibitors: Design Concepts
By Weiping Zheng

Published by the Royal Society of Chemistry, www.rsc.org

CHAPTER 1

Catalytic Mechanism-based Design

1.1 Mode of Working

Enzymes use different mechanistic strategies to fulfill their mission as biocatalysts to accelerate chemical reactions. Currently, the term "enzyme" refers to a protein-based biocatalyst or a nucleic acid-based biocatalyst (primarily the catalytic ribonucleic acid). While these two types of enzymes are of two different chemical classes, they rely on largely the same catalytic mechanisms to accelerate a chemical reaction.[1]

The catalytic mechanism of an enzyme-catalyzed reaction (or enzymatic reaction) can be categorized as the kinetic mechanism and the chemical mechanism. The kinetic mechanism is concerned with the kinetic aspect of an enzymatic reaction, which can also be regarded as the physical and macroscopic aspect of an enzymatic reaction, *i.e.* the order of substrate binding and product release and the identity of the enzyme–substrate complex in which a chemical transformation *en route* from substrate to product occurs. The chemical mechanism is concerned with the chemical strategy employed by an enzyme to accelerate the chemical transformation from substrate to product within the enzyme–substrate complex. Both kinetic and chemical mechanisms can be exploited profitably in the design of the catalytic mechanism-based inhibitor for an enzymatic reaction. Since every enzymatic reaction has to rely on a catalytic mechanism to accelerate the reaction, the catalytic mechanism-based design strategy is applicable to all types of enzymatic reactions.

If we can design a compound whose chemical structure is similar to that of a substrate of an enzymatic reaction, so that the enzyme can take it as an alternate substrate and process it as during normal catalysis, yet with the formation of an enzyme-bound catalytic intermediate/product that occupies tightly the enzyme active site *via* non-covalent binding or a covalent bond

Active Site-directed Enzyme Inhibitors: Design Concepts
By Weiping Zheng

Published by the Royal Society of Chemistry, www.rsc.org

between the catalytic intermediate/product and enzyme active site, such a compound is then called a catalytic mechanism-based inhibitor for this enzymatic reaction. Whether or not there is a covalent binding for the inhibitor-derived catalytic intermediate/product at enzyme active site, the enzymatic reaction is to be inhibited irreversibly, leading to a time-dependent inactivation.[2,3]

1.2 Applications

Enzymatic reactions can be classified into six different types based on the types of the chemical reactions they catalyze, namely (i) transferases catalyzing functional group transfer reactions; (ii) oxidoreductases catalyzing oxidation–reduction reactions; (iii) hydrolases catalyzing hydrolytic reactions; (iv) lyases catalyzing group elimination reactions; (v) isomerases catalyzing isomerization reactions; and (vi) ligases catalyzing ligation (*e.g.* C–C bond formation) reactions.[4] Notable examples and the current status of development of the catalytic mechanism-based inhibitors for each of these six types of the enzymatic reactions will be elaborated herein, to showcase the elegancy of this design concept. It is worth noting that, in addition to being potential therapeutic agents, these inhibitory compounds could also be employed as chemical tools, helping to dissect the biology and pharmacology of the enzymatic reactions concerned.

1.2.1 Applications with Transferases

In this section, notable examples of the transferase-catalyzed reactions are employed to illustrate how the catalytic mechanism-based inhibitors have been elegantly developed for this class of enzymatic reactions. We can see that either covalent or non-covalent catalytic mechanism-based inhibition can be achieved with a transferase-catalyzed reaction: in the former case, there is a generation of an inhibitor-derived catalytic intermediate or product able to bind covalently at enzyme active site; whereas in the latter case, the inhibitor-derived catalytic intermediate or product is unable to bind covalently at enzyme active site, yet it has a strong non-covalent binding interaction with the enzyme active site. Therefore, covalent and non-covalent inhibition are both able to prevent an enzyme active site from binding and processing the normal substrate in a time-dependent manner, thereby leading to an irreversible inhibition of the normal enzymatic catalysis. Moreover, in the case of covalent inhibition, the cofactor (*e.g.* coenzyme) required for enzymatic catalysis can also be the interaction partner of the inhibitor-derived catalytic intermediate or product at the enzyme active site.

1.2.1.1 Catalytic Mechanism-based Inhibitors for the Sirtuin-catalyzed Deacylation Reaction

"Sirtuin" is the name of a family of β-nicotinamide adenine dinucleotide (β-NAD^+ or just NAD^+)-dependent protein N^{ε}-acyl-lysine deacylases whose enzymatic action serves to counteract the post-translational lysine side chain N^{ε}-acylation and which are conserved in organisms from the three evolutionary life forms (*i.e.* bacteria, archaea, and eukarya).[5–7] In mammals (including humans), seven sirtuin isoforms (*i.e.* SIRT1–7) have been identified. Figure 1.1 depicts the latest version of the chemical mechanism proposed for the deacylation reaction catalyzed by sirtuin family members.[6]

It should be noted that even though these enzymes are deacylases and water serves as a substrate at the last stage on the deacylation reaction coordinate, the net outcome of this enzymatic reaction is the transfer of the acyl group from the N^{ε}-acyl-lysine substrate onto coenzyme NAD^+ with concomitant cleavage of the nicotinamide moiety from NAD^+, ultimately leading to the formation of the deacylated product and the acylated ADP-ribose (*i.e.* 2′-*O*-acyl-ADP-ribose (2′-*O*-AADPR)); moreover, the first chemical step is actually an ADP-ribosylation of the N^{ε}-acyl-lysine side chain amide oxygen. Therefore, the sirtuin-catalyzed deacylation is a functional group transfer reaction.

The sirtuin-catalyzed deacylation reaction has been found to be important in regulating a variety of vital life processes such as gene transcription, metabolism, DNA damage repair, neurodegeneration, and aging. This enzymatic reaction has thus been regarded as a potential therapeutic target for developing novel therapeutics for treating human diseases such as cancer and metabolic and neurodegenerative diseases. As a result, chemical modulators of the sirtuin-catalyzed deacylation reaction have been actively pursued.[7]

Concerning the inhibitor development, more than one type of the catalytic mechanism-based inhibitors have been developed for the sirtuin-catalyzed deacylation reaction, with each type harboring one unique type of so-called "catalytic mechanism-based sirtuin inhibitory warhead" that confers upon the inhibitor molecule a catalytic mechanism-based sirtuin inhibition. Of note, the prototypical and up-to-date most powerful type of the catalytic mechanism-based sirtuin inhibitory warheads are those based on N^{ε}-thioacyl-lysines which are one type of the close structural analogs of N^{ε}-acyl-lysine found in a normal substrate of the sirtuin deacylation reaction.

Figure 1.2 depicts the sirtuin-catalyzed processing of an inhibitor molecule harboring the catalytic mechanism-based sirtuin inhibitory warhead N^{ε}-thioacyl-lysine.[7,8] Such an inhibitor can be taken as an alternate substrate by a sirtuin and be processed as depicted with the formation of a longer-lived (or stalled) catalytic intermediate (I, II, or III) that non-covalently binds at the sirtuin active site. Such a stalled intermediate presumably would be a tight-binding bi-substrate analog inhibitor of the sirtuin that catalyzes its formation, since its two structural parts are derived respectively from the thioacyl-lysine inhibitor and coenzyme NAD^+. The sirtuin-catalyzed formation of such a stalled catalytic intermediate could be viewed as a perfect way of producing a tight-binding bi-substrate analog inhibitor for itself. Tight-binding behavior of such an inhibiting species would have contributed to the experimentally observed time-dependent inhibition profile and would

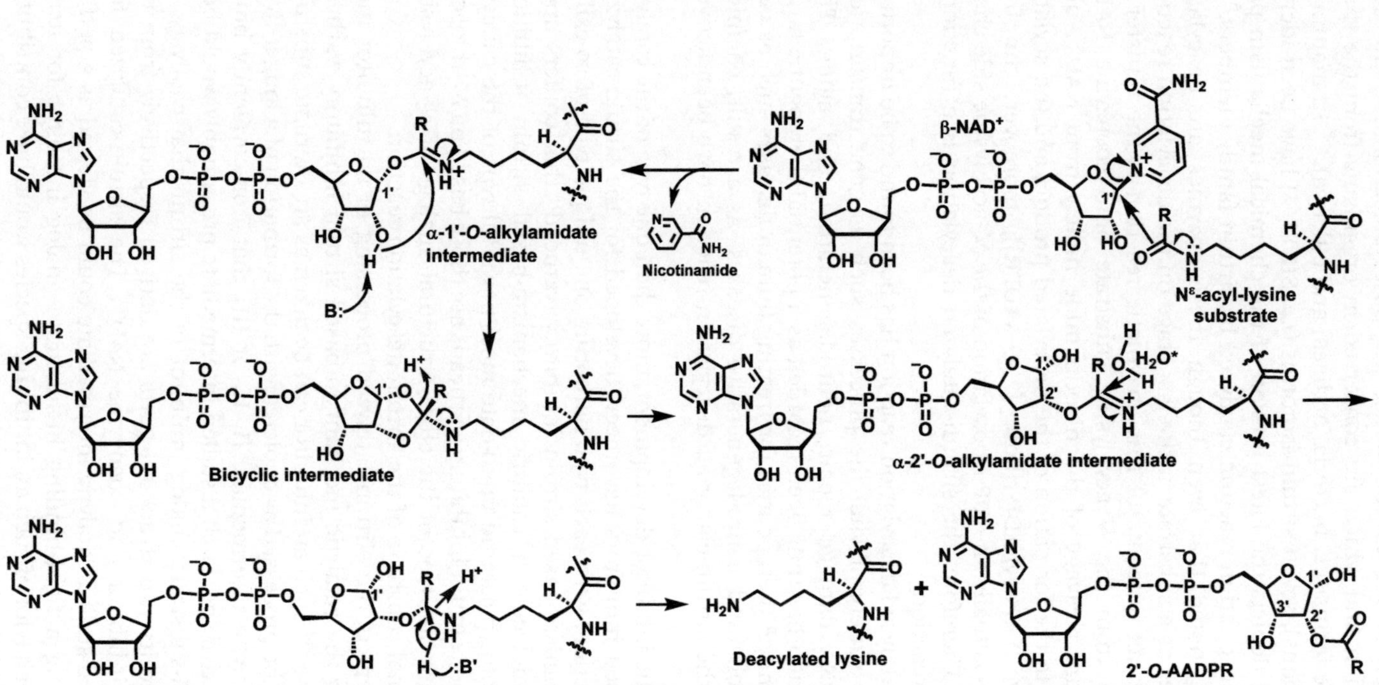

Figure 1.1 The latest version of the chemical mechanism proposed for the sirtuin-catalyzed deacylation reaction. The catalytically removable acyl group (*i.e.* R–(C=O)–) could be acetyl or bulkier ones such as succinyl and myristoyl; B: and :B′, general bases from the sirtuin active site; the asterisk in "H_2O*" denotes that this water molecule may be activated by a sirtuin active site general base or by a specific base in bulk solution; the proton (H^+) sources may be a sirtuin active site general acid or bulk solution.

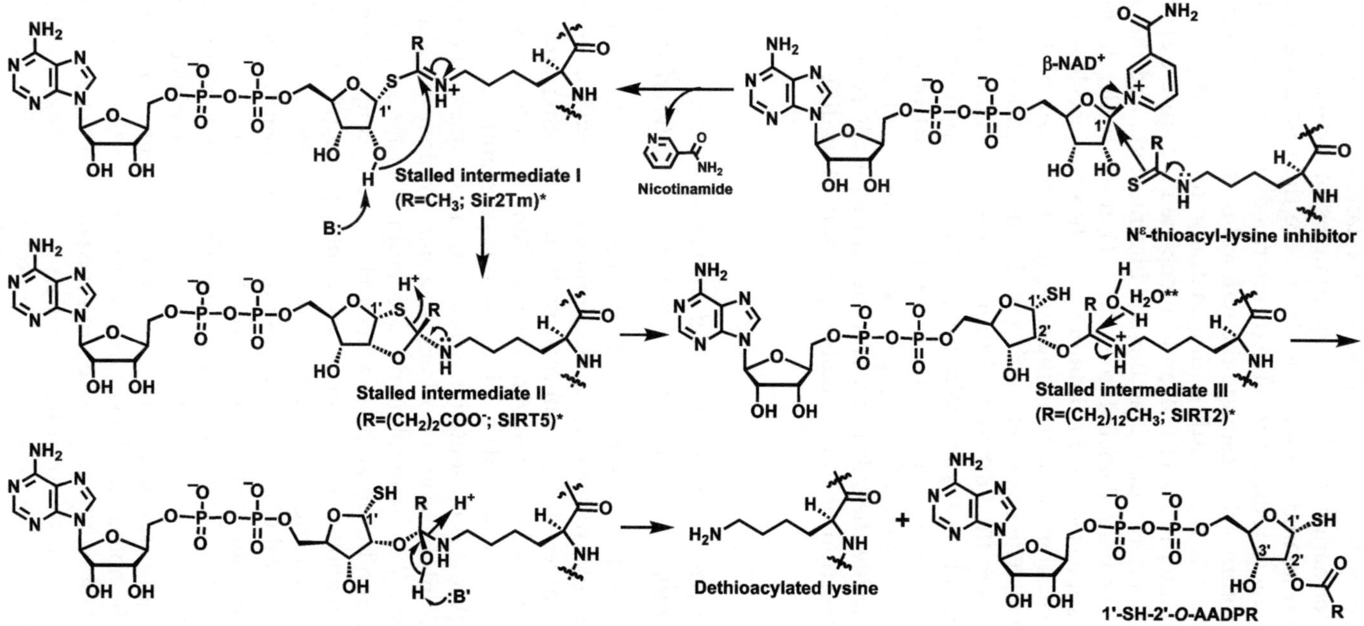

Figure 1.2 The sirtuin-catalyzed processing of an inhibitor harboring the catalytic mechanism-based sirtuin inhibitory warhead N^{ε}-thioacyl-lysine. This scheme is essentially same as that shown in Figure 1.1. The single asterisk sign denotes that the indicated sirtuin isoform and the N^{ε}-thioacyl-lysine inhibitor with the indicated R group are those employed in a study that furnished the first structural view of the relevant stalled intermediate; B: and :B′, general bases from the sirtuin active site; the double asterisk sign in "H_2O**" denotes that this water molecule may be activated by a sirtuin active site general base or by a specific base in bulk solution; the proton (H^+) sources may be a sirtuin active site general acid or bulk solution.

have been made possible considering the sequential ternary complex kinetic mechanism obeyed by the sirtuin-catalyzed deacylation reaction (*i.e.* there is an obligatory formation of the ternary complex of a sirtuin with the N^{ε}-acyl-lysine substrate and coenzyme NAD^+ before the first chemical step occurs). It is amazing that different catalytic intermediates can be stalled by different sirtuin isoforms that catalyze the removal of different acyl groups, as shown in Figure 1.2. It is worth noting that, due to the close structural similarity between intermediates I, II, and III (depicted in Figure 1.2) and the α-1′-*O*-alkylamidate intermediate, the bicyclic intermediate, and the α-2′-*O*-alkylamidate intermediate (depicted in Figure 1.1), respectively, the use the catalytic mechanism-based sirtuin inhibitory warhead thioacyl-lysines has also been instrumental in elucidating the catalytic mechanism of the sirtuin deacylation reaction. The observed stalling of the afore-mentioned catalytic intermediates I, II, and III at different sirtuin active sites could be accounted for by the differentially changed positioning of the corresponding catalytic intermediates (depicted in Figure 1.1) at different sirtuin active sites during normal enzymatic catalysis following the isosteric replacement of sulfur for oxygen of the N^{ε}-acyl-lysine substrate, and the consequently stalled further chemical transformation of intermediates I, II, and III at different sirtuin active sites.

1.2.1.2 A Catalytic Mechanism-based Inhibitor for the γ-Aminobutyric Acid Aminotransferase-catalyzed Reaction

As depicted in Figure 1.2, the inhibitors harboring the catalytic mechanism-based sirtuin inhibitory warhead N^{ε}-thioacyl-lysines work *via* a non-covalent inhibition of the target sirtuin isoforms, with the inhibiting species presumably being the covalent conjugate derived from the thioacyl-lysine inhibitor and the coenzyme NAD^+, which is non-covalently bound to the sirtuin active site. What is presented here is a further example of the catalytic mechanism-based inhibitor also working *via* a non-covalent inhibition of a transferase enzyme (*i.e.* γ-aminobutyric acid aminotransferase (GABA-AT)) with the putative inhibiting species being the covalent conjugate derived from the inhibitor and coenzyme pyridoxal 5′-phosphate (PLP). The inhibition of this GABA-AT-catalyzed reaction has been proven to be beneficial for treating neurological disorders such as seizures by elevating the level of the inhibitory neurotransmitter GABA in the human brain.[9]

Figure 1.3 depicts the chemical mechanism of the PLP-dependent reaction catalyzed by GABA-AT, *i.e.* conversion of GABA to succinic semialdehyde with the concomitant transformation of PLP to pyridoxamine 5′-phosphate (PMP).[9] It should be noted that the key mechanistic step is the abstraction of the γ-proton of GABA by a general base from the GABA-AT active site, which presumably is driven by the pyridinium cation (electron sink) on PLP. The subsequent electron pushing affords a new protonated Schiff base engaging the γ-carbon of GABA instead of the original benzylic carbon of PLP,

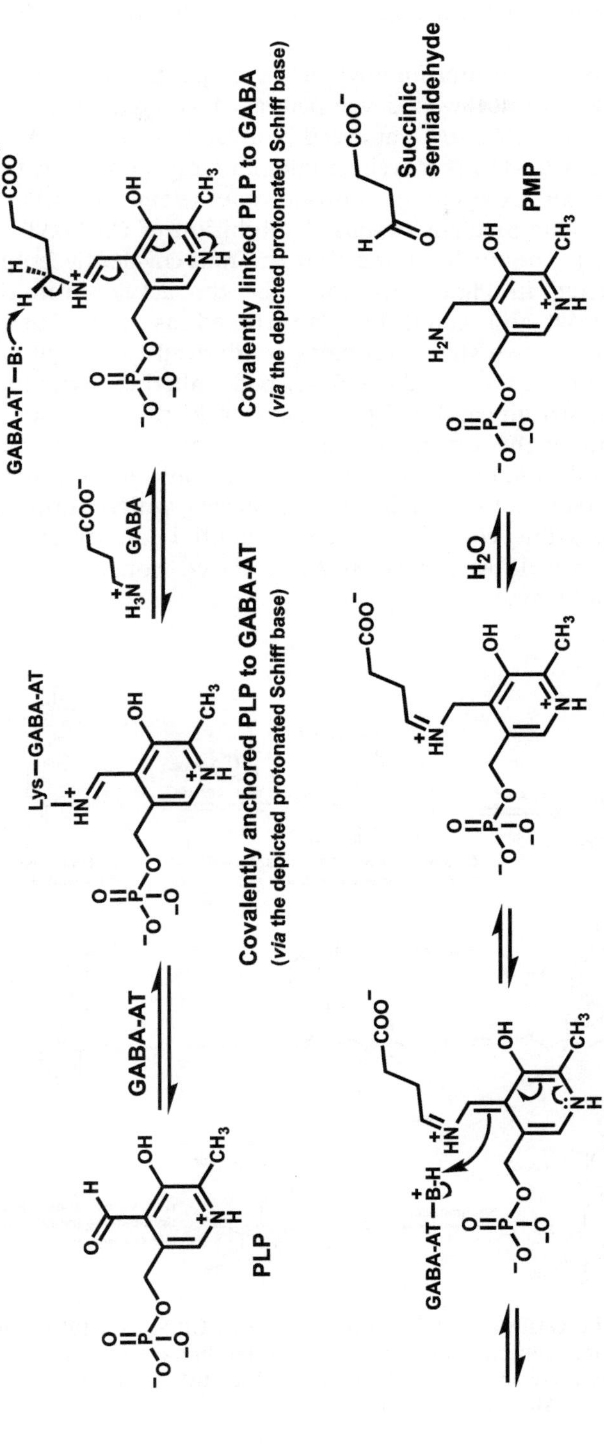

Figure 1.3 The proposed chemical mechanism for the GABA-AT-catalyzed conversion of GABA to succinic semialdehyde with the concomitant transformation of coenzyme PTP to PMP. Lys and B: respectively refer to a lysine residue and a general base from GABA-AT active site.

the hydrolysis of which furnishes the two end products of this enzymatic reaction.

(1*S*,3*S*)-3-Amino-4-difluoromethylene-1-cyclopentanoic acid (CPP-115) depicted in Figure 1.4 was developed as a catalytic mechanism-based inhibitor for the GABA-AT-catalyzed conversion of GABA to succinic semialdehyde.[9] CPP-115 is a cyclic analog of GABA, however, the judicious placement of an exocyclic *gem*-difluoromethylene on CPP-115 makes possible a catalytic mechanism-based inhibition of GABA-AT, as depicted in Figure 1.4. It should be noted that the non-covalently bound catalytic product depicted in Figure 1.4 could be the actual inhibiting species against GABA-AT and could be also viewed as a bi-substrate analog inhibitor, since its two structural parts are derived from CPP-115 and PLP, a scenario analogous to the above-described stalled catalytic intermediates formed by the sirtuin-catalyzed processing of N^{ε}-thioacyl-lysine inhibitors. Also analogous to the sirtuin scenario, the experimentally observed time-dependent inhibition of GABA-AT by CPP-115 could have resulted from the tight-binding behavior of this inhibiting species, which in turn would have been made possible considering the sequential ternary complex kinetic mechanism obeyed by the GABA-AT-catalyzed conversion of GABA to succinic semialdehyde.

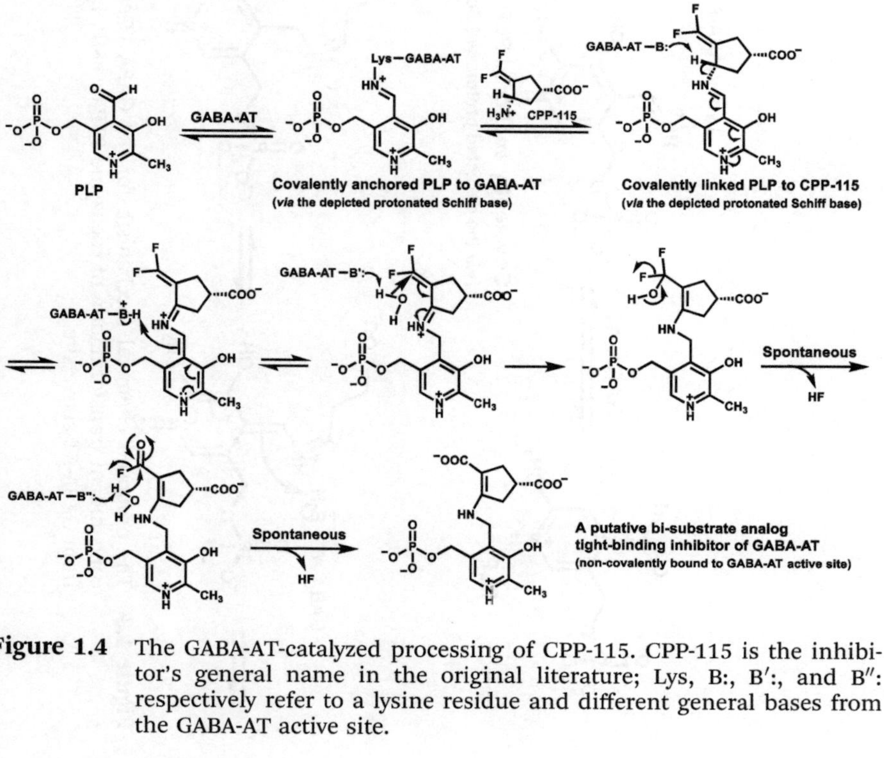

Figure 1.4 The GABA-AT-catalyzed processing of CPP-115. CPP-115 is the inhibitor's general name in the original literature; Lys, B:, B′:, and B″: respectively refer to a lysine residue and different general bases from the GABA-AT active site.

1.2.1.3 *A Catalytic Mechanism-based Inhibitor for the DOT1L-catalyzed Histone H3K79 Methylation Reaction*

As another example of catalytic mechanism-based non-covalent inhibition of an enzymatic transfer reaction, compound **4** (the compound numbering in the original literature), an *S*-(5′-adenosyl)-L-methionine (SAM) analog shown in Figure 1.5, was found to strongly inhibit the DOT1L-catalyzed SAM-dependent methylation of the side chain ε-amino group of histone H3K79.[10] It should be noted that DOT1L is a histone methyltransferase with specificity toward H3K79 located in the nucleosomal core octamer and the DOT1L-catalyzed H3K79 side chain methylation was shown to be important in regulating cell differentiation and promoting the development of acute leukemia with mixed-lineage leukemia gene translocations.[10]

As depicted in Figure 1.6, compound **4** can be processed by DOT1L in the same manner as SAM, but with the formation of a covalent conjugate which presumably would behave as the true inhibiting bi-substrate analog inhibitor non-covalently and tightly bound to the DOT1L active site.[10] It should be noted that, even though the 5′-tertiary amine of compound **4** is predominantly protonated in bulk solution under physiological pH, the pK_b (negative logarithm of the base ionization constant) of this amine could be enhanced within the DOT1L active site due to the more hydrophobic DOT1L active site microenvironment than that in bulk solution so as to enhance the free base proportion of the 5′-tertiary amine of compound **4**, consequently promoting the formation of the depicted aziridinium, a mimic of the sulfonium of SAM.

1.2.1.4 *A Catalytic Mechanism-based Inhibitor for the Glutathione Transferase-catalyzed Reaction*

It should be noted that all the catalytic mechanism-based inhibitors described above for different transferase-catalyzed reactions not only are non-covalent inhibitors, but also invoke the participation of different coenzymes (*e.g.* NAD^+ and PLP) as part of a true inhibiting bi-substrate analog inhibitor. Presented in Figure 1.7(B) is an example of the catalytic mechanism-based inhibitor whose mode of inhibition does not invoke the participation of a coenzyme, namely that for the glutathione *S*-transferase (GST)-catalyzed

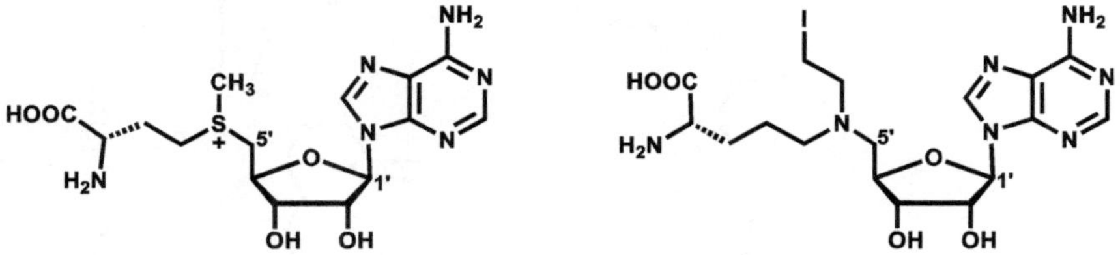

Figure 1.5 The chemical structures of SAM and compound **4**.

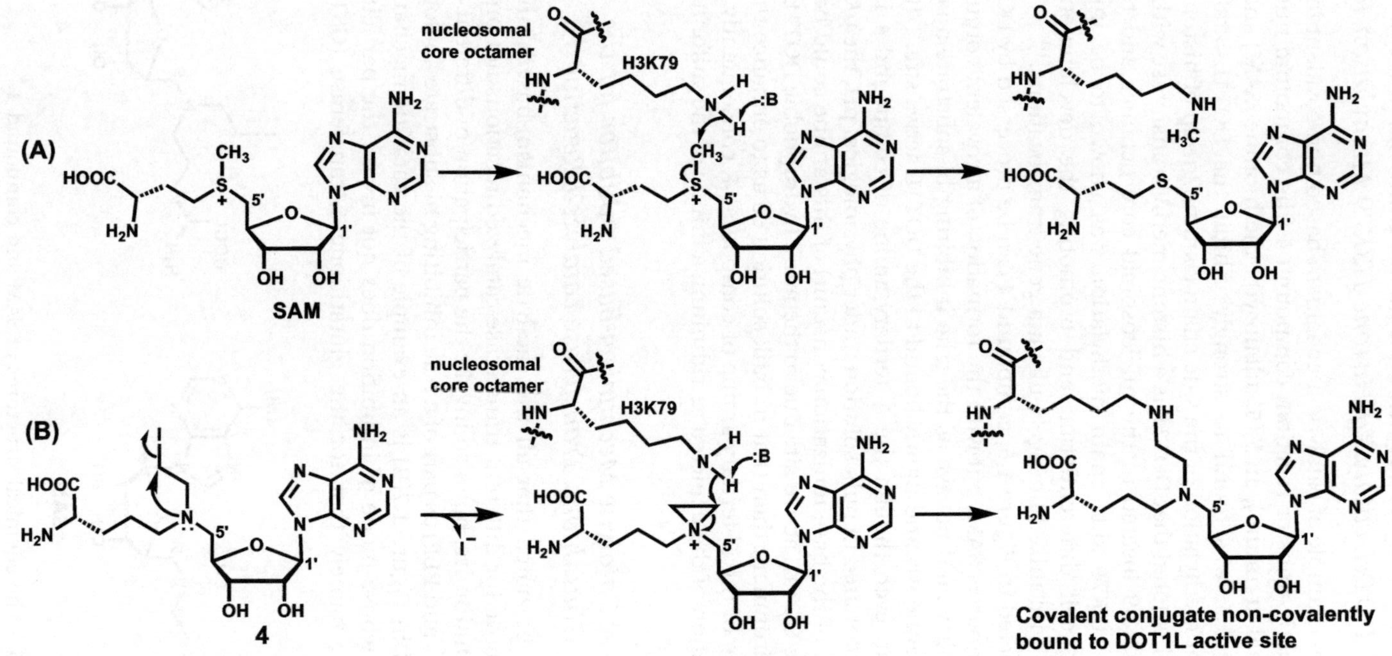

Figure 1.6 The DOT1L-catalyzed processing of (A) SAM and (B) compound **4**. B:, a general base at the DOT1L active site.

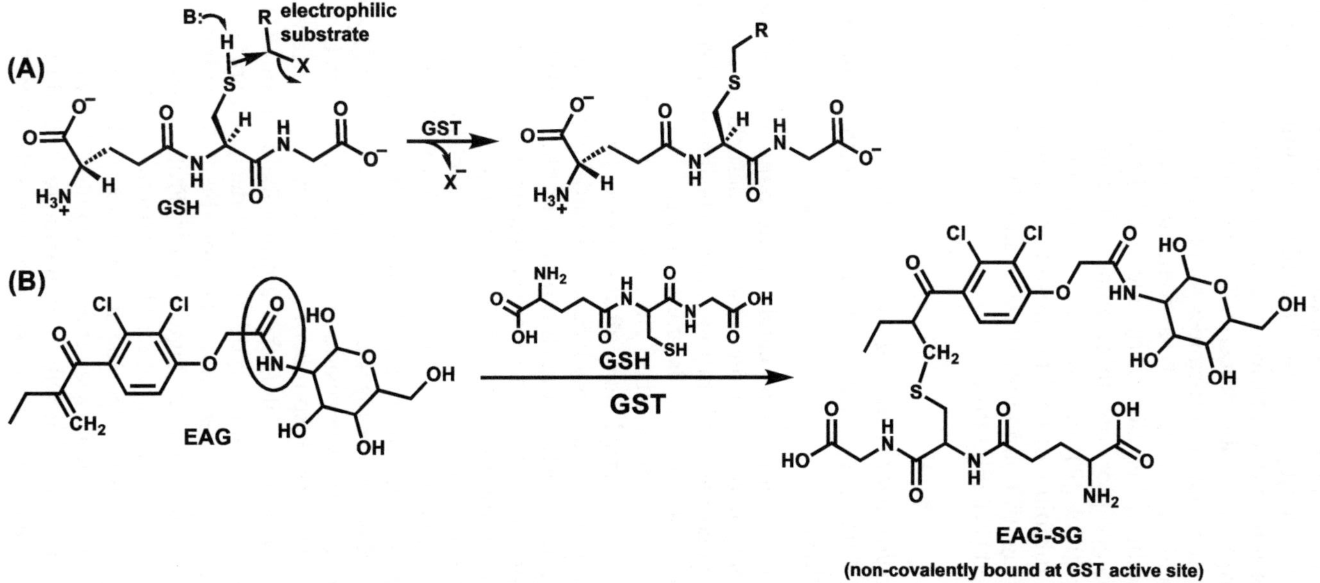

Figure 1.7 (A) The GST-catalyzed nucleophilic substitution (one type of conjugation reaction) between GSH and an electrophilic substrate. B:, a general base at the GST active site. (B) The GST-catalyzed conjugate addition between EAG and GSH with the formation of a bi-substrate analog inhibitor (*i.e.* EAG-SG) non-covalently bound at the GST active site. The amide linkage in EAG is circled; the stereochemistry in all the stereogenic centers is not explicitly indicated in this scheme.

glutathione (GSH) conjugation reaction (*e.g.* nucleophilic substitution) with an electrophilic substrate depicted in Figure 1.7(A).[11] Of note, the GST-catalyzed GSH conjugation reaction plays an important role in xenobiotic detoxification and oxidative stress, properties contributing to its demonstrated exploitation by cancer cells. Therefore, GST inhibitors might be exploited for developing anti-cancer therapeutics.[11]

As depicted in Figure 1.7(B), GST was found to be also capable of catalyzing the conjugate addition reaction of the electrophilic EAG (an amide formed from the Michael acceptor-containing ethacrynic acid and glucosamine) with the nucleophilic thiol (SH) of GSH, and the reaction product EAG-SG was found to be an inhibitor of GST by presumably behaving as a bi-substrate analog inhibitor tightly bound to GST active site. Therefore, EAG is a catalytic mechanism-based GST inhibitor.

1.2.1.5 Catalytic Mechanism-based Inhibitors for the Human DNA Cytosine-5 Methyltransferase-catalyzed Methylation Reaction

In addition to the examples of catalytic mechanism-based non-covalent inhibition of enzymatic transfer reactions described above, catalytic mechanism-based covalent inhibition is also possible for certain enzymatic transfer reactions, the following is one example.

Human DNA cytosine-5 methyltransferase (DNMT) catalyzes the SAM-dependent C-5 methylation of the nucleobase cytosine with the formation of the so-called 5th nucleobase in CpG islands of genomic DNA, *i.e.* 5-methylcytosine (mC).[12] The inhibition of the DNMT-catalyzed formation of mC holds promise as a cancer treatment strategy, since tumor cells tend to harbor aberrant DNA methylation patterns.[12]

Figure 1.8 presents the proposed chemical mechanism for the DNMT-catalyzed methylation reaction.[12] The key feature of this mechanism is the pre-anchoring of the cytosine of an oligodeoxyribonucleotide at C-6 onto a DNMT active site Cys residue for the subsequent catalytic methylation at C-5 of cytosine, eventually the Cys residue gets liberated back to the initial state at the DNMT active site. Inspired by this mechanism, a few oligodeoxyribonucleotides harboring different 2-amino-4-halopyridine-C-nucleosides (mimicking the C-6 halogenated cytidine) were designed and were found to be catalytic mechanism-based inhibitors for the DNMT-catalyzed methylation reaction, as depicted in Figure 1.9.[12] From what is depicted in Figures 1.8 and 1.9, the binding of an oligodeoxyribonucleotide harboring a 2-amino-4-halopyridine-C-nucleoside at the DNMT active site could position the cytosine analog in such a way that the bound SAM would just serve a structural role rather than participating in catalysis. In addition, from what is depicted in Figure 1.9, the net reaction between the DNMT active site Cys residue and a 2-amino-4-halopyridine-C-nucleoside would be a nucleophilic aromatic substitution (S_NAr) reaction, which would be consistent with the

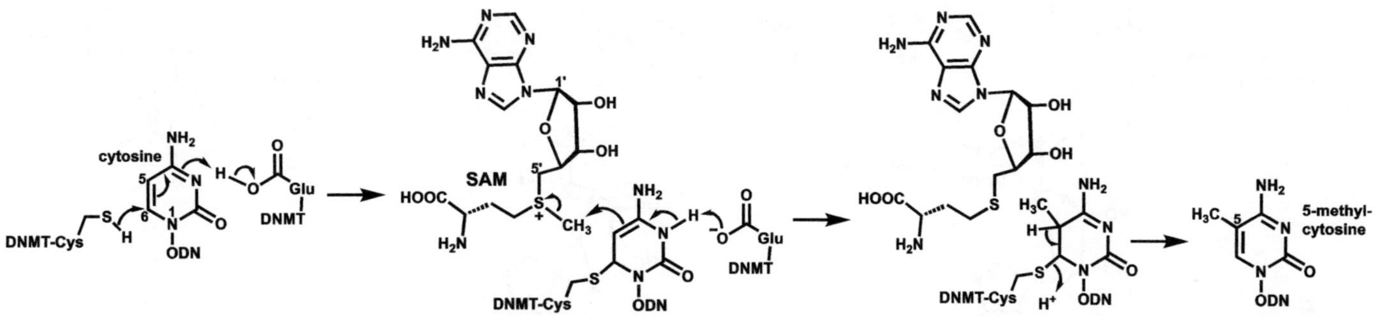

Figure 1.8 The proposed chemical mechanism for the DNMT-catalyzed SAM-dependent cytosine C-5 methylation. Cys, cysteine; Glu, glutamic acid; ODN, oligodeoxyribonucleotide.

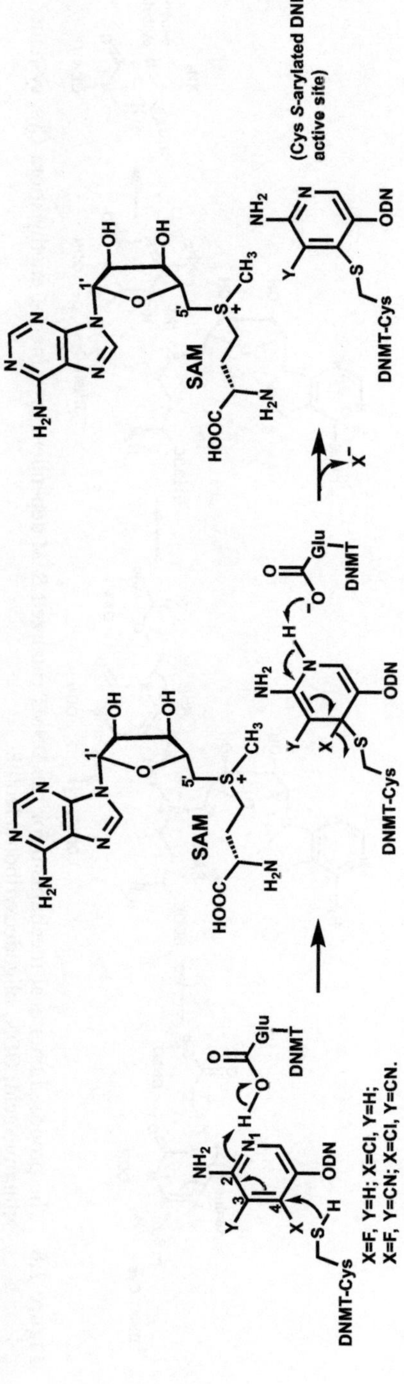

Figure 1.9 The DNMT-catalyzed processing of the oligodeoxyribonucleotides harboring different 2-amino-4-halopyridine-C-nucleosides. Cys, cysteine; Glu, glutamic acid; ODN, oligodeoxyribonucleotide.

observed faster reaction with X = F in the analog than that with X = Cl (less electron-withdrawing than F), and with that the introduction of the electron-withdrawing CN at the C-3 position in the analog (*i.e.* Y = CN) was observed to promote the reaction.

1.2.2 Applications with Oxidoreductases

Oxidoreductases include oxidases, reductases, and dehydrogenases and catalyze a variety of oxidation or reduction reactions involved in various life processes. The catalytic mechanism-based inhibitors have been elegantly developed for various oxidoreductase-catalyzed oxidation–reduction reactions, as illustrated herein with notable examples. As for the scenario with the catalytic mechanism-based inhibitors for the transferase-catalyzed reactions described above, a time-dependent covalent or non-covalent irreversible inhibition can also be achieved with a catalytic mechanism-based inhibitor for the oxidoreductase-catalyzed reaction.

1.2.2.1 A Catalytic Mechanism-based Inhibitor for the Lysine-specific Demethylase 1-Catalyzed Demethylation Reaction

The lysine-specific demethylase 1 (LSD1) is a transcriptional repressor and a nuclear amine oxidase homolog able to catalyze the N^{ε}-methyl hydroxylation on the N^{ε}-mono- and -dimethylated lysine side chains in various proteins, such as the lysine 4 N^{ε}-methylated histone H3 protein. The resulting hemiacetal-like intermediate would then spontaneously collapse and give rise to formaldehyde and the demethylated protein species.[13] As shown in Figure 1.10, the first mechanistic step is the flavin adenine dinucleotide (FAD)-dependent oxidation of the N^{ε}-methyl functionality with the formation of the depicted iminium, whose hydrolysis would then afford the hemiacetal-like intermediate.[14]

Based on this mechanistic scheme, a peptide corresponding to the first 21 amino acids of the histone H3 protein in which the methylated lysine side chain was replaced with N^{ε}-propargyl-lysine, and the resulting compound (*i.e.* H3-K4(propargyl)(1–21) depicted in Figure 1.11) was found to be an irreversible catalytic mechanism-based inhibitor for the LSD1-catalyzed demethylation reaction, consistent with a time-dependent loss of LSD1 enzymatic activity following inhibitor treatment.[13]

As depicted in Figure 1.11, H3-K4(propargyl)(1–21) can be also taken up by LSD1 as a substrate and processed, however, the resulting iminium functionality is part of the depicted 1,4-unsaturated system that can serve as the Michael acceptor during the 1,4-conjuated addition with the N5 of $FADH_2$, leading to an irreversible inhibition of the LSD1-catalyzed FAD-dependent demethylation reaction.[14]

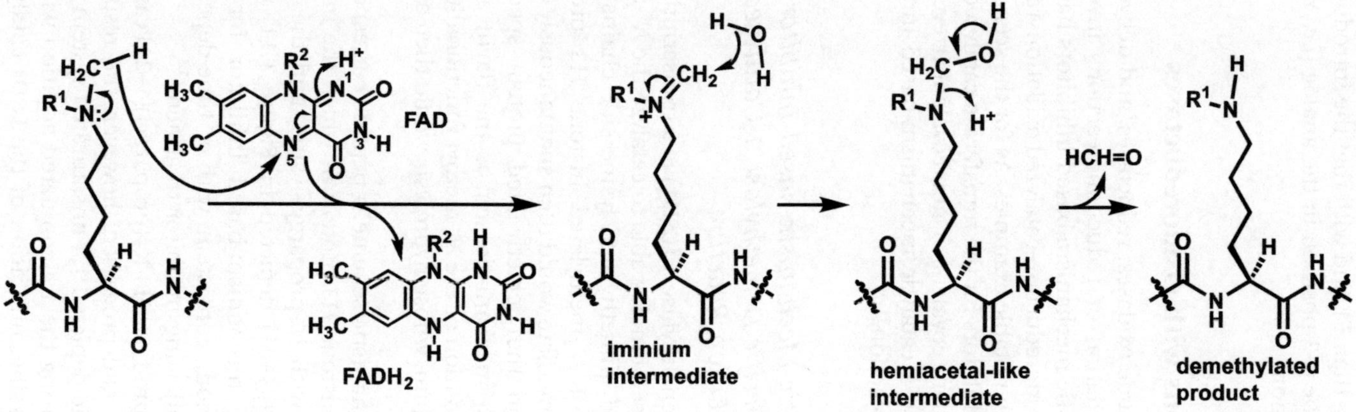

Figure 1.10 The proposed chemical mechanism for the LSD1-catalyzed FAD-dependent demethylation reaction. R^1, CH_3 or H; R^2, ribosyl adenine dinucleotide. Please see Figure 5.11 in this book for the full chemical structures of FAD and $FADH_2$.

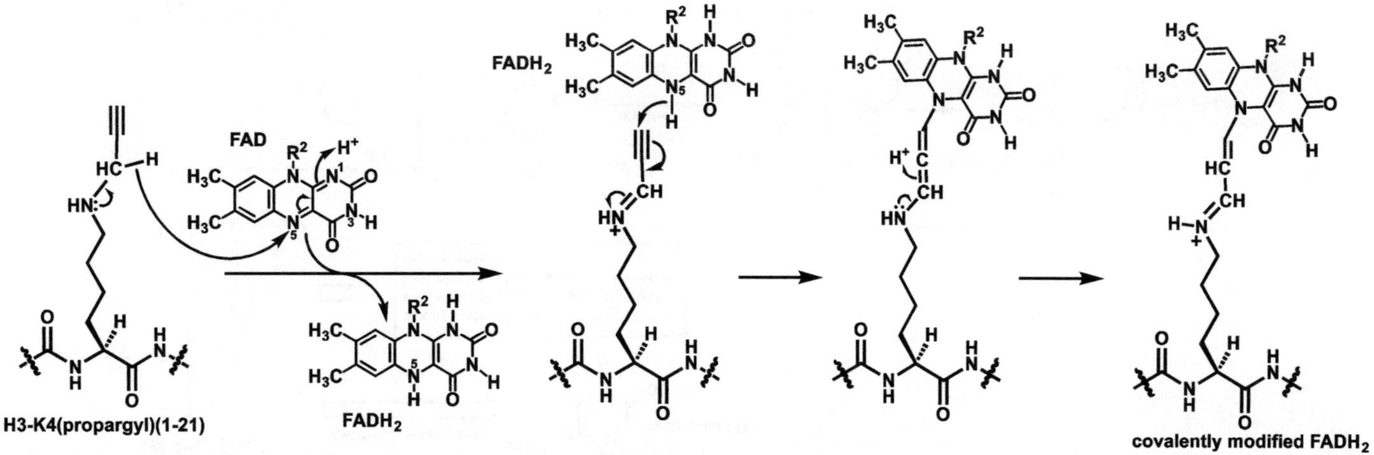

Figure 1.11 The LSD1-catalyzed FAD-dependent processing of H3-K4(propargyl)(1–21) leading to the indicated covalent modification of the coenzyme $FADH_2$. R^2, ribosyl adenine dinucleotide. Please see Figure 5.11 in this book for the full chemical structures of FAD and $FADH_2$.

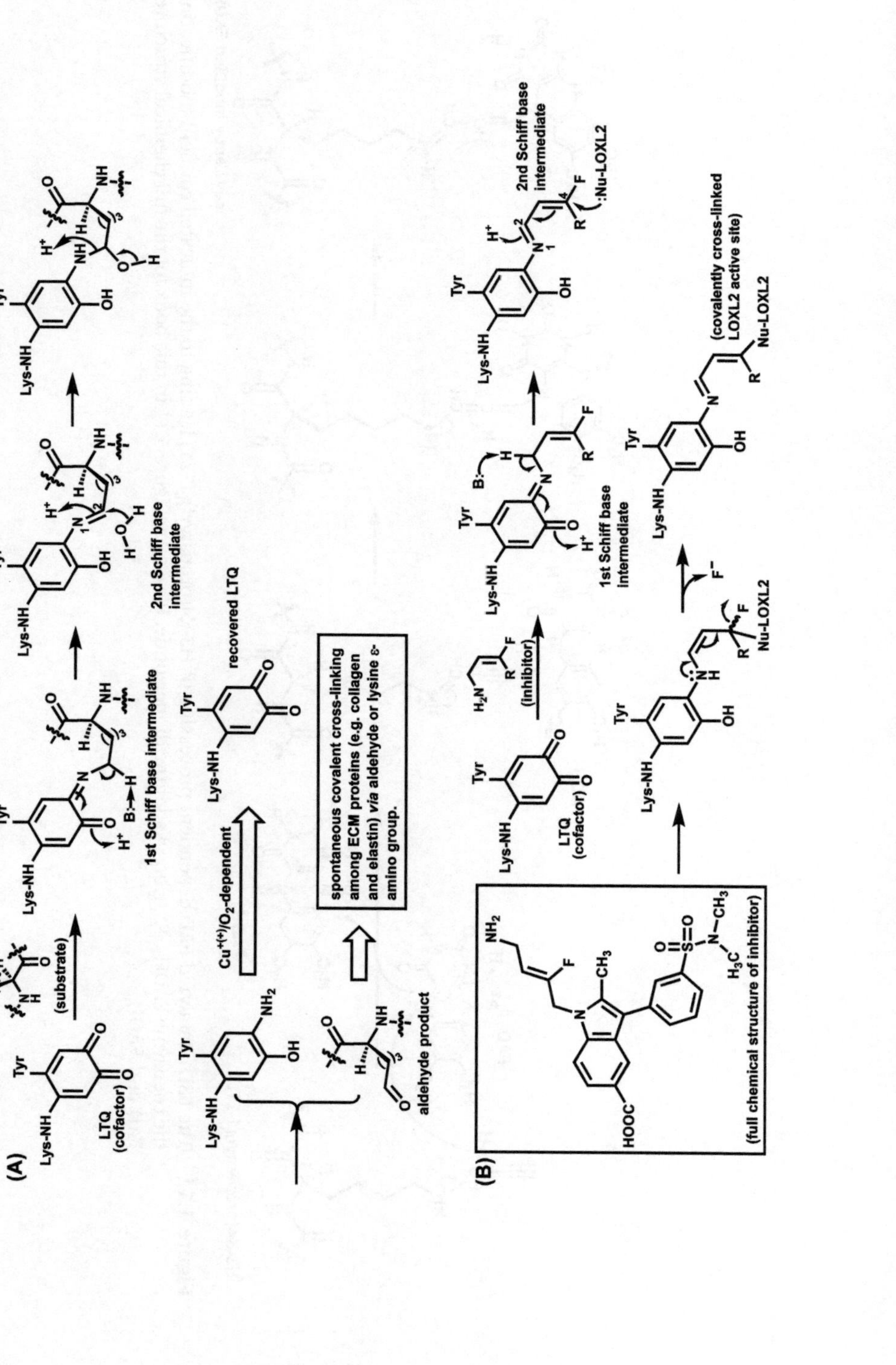
(A)
Lys-NH
Tyr
LTQ
(cofactor)
(substrate)
1st Schiff base intermediate
2nd Schiff base
intermediate
Cu(+)/O2-dependent
recovered LTQ
aldehyde product
spontaneous covalent cross-linking among ECM proteins (e.g. collagen and elastin) via aldehyde or lysine ε-amino group.
(B)
(full chemical structure of inhibitor)
(inhibitor)
1st Schiff base
intermediate
2nd Schiff base
intermediate
:Nu-LOXL2
Nu-LOXL2
(covalently cross-linked
LOXL2 active site)

1.2.2.2 A Catalytic Mechanism-based Inhibitor for the Lysyl Oxidase-like 2-Catalyzed Cross-linking Reaction

It should be noted that the catalytic mechanism-based inhibitor for the LSD1-catayzed FAD-dependent demethylation reaction described above ought to exhibit a non-covalent inhibition, since the cofactor FAD is non-covalently bound to LSD1, however, a couple of examples of the catalytic mechanism-based inhibitors for the oxidoreductase-catalyzed reactions described in this section and the following would be able to achieve a covalent inhibition, since the catalytic cofactors are bound covalently to the respective oxidoreductases.

Lysyl oxidase-like 2 (LOXL2) is a $Cu^{+(+)}$-dependent amine oxidase able to catalyze the oxidative ε-deamination of lysine residues in extracellular matrix (ECM) proteins such as collagen and elastin, and the resultant aldehyde product is able to initiate the spontaneous covalent cross-linking among such ECM proteins *via* aldehyde or lysine ε-amino group.[15,16] Figure 1.12(A) shows the proposed chemical mechanism for the LOXL2-catalyzed lysine tyrosylquinone (LTQ)-dependent ε-deamination of a lysine residue in a substrate *via* the two depicted Schiff base intermediates, which would be followed by the final recovery of the cofactor LTQ at the LOXL2 active site *via* a $Cu^{+(+)}$/O_2-dependent process.[15] The pharmacological inhibition of the LOXL2-catalyzed reaction therefore represents a therapeutic target for fibrotic diseases. Figure 1.12(B) depicts the proposed mode of catalytic mechanism-based inhibition of the LOXL2-catalyzed reaction by a fluoroallylamine-based inhibitor.[16] The key mechanistic step in this proposed scheme is the nucleophilic 1,4-addition by a LOXL2 active site nucleophile instead of the nucleophilic 1,2-addition by a water molecule (like that during normal enzymatic catalysis shown in Figure 1.12(A)) on the second Schiff base intermediate, ultimately leading to the covalently cross-linked LOXL2 active site.

1.2.2.3 A Catalytic Mechanism-based Inhibitor for the Human Persulfide Dioxygenase-catalyzed Dioxygenation Reaction

The mononuclear metalloenzyme persulfide dioxygenase (PDO) employs a covalently bound non-heme iron center as its catalytic cofactor for the

Figure 1.12 (A) The proposed chemical mechanism for the LOXL2-catalyzed LTQ-dependent ε-deamination of a lysine residue in a substrate *via* the two depicted Schiff base intermediates, which would be followed by the final recovery of the cofactor LTQ at the LOXL2 active site *via* a $Cu^{+(+)}$/O_2-dependent process. B:, a general base at the LOXL2 active site. Note: the cofactor LTQ is originally formed on LOXL2 from lysine (Lys) and tyrosine (Tyr). (B) The proposed catalytic mechanism-based inhibitory mode of LOXL2 by a fluoroallylamine-based inhibitor. :Nu-LOXL2, a nucleophile at the LOXL2 active site.

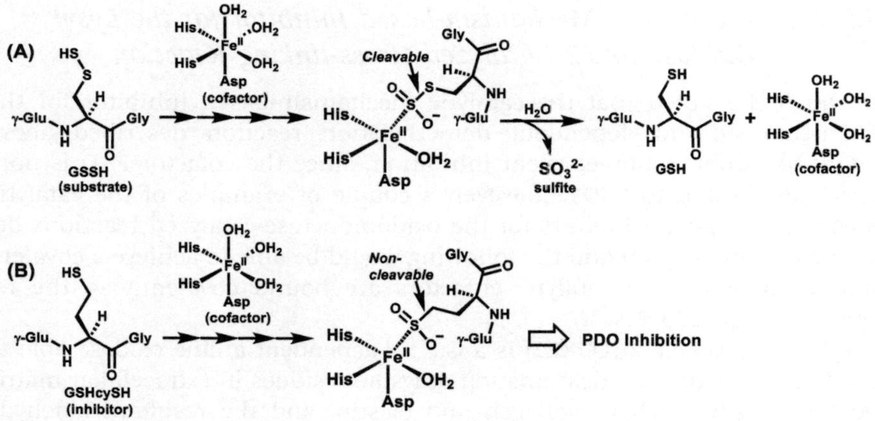

Figure 1.13 (A) The proposed chemical mechanism for the PDO-catalyzed conversion of GSSH to sulfite (SO_3^{2-}) and GSH. (B) The proposed chemical mechanism for the catalytic mechanism-based inhibition of the PDO-catalyzed reaction by GSHcySH. The cleavable S–S bond and the non-cleavable S–C bond are indicated.

conversion of glutathione persulfide (GSSH) to sulfite (SO_3^{2-}) and GSH.[17] The PDO-catalyzed reaction plays an important role in regulating the cellular concentration of the signaling molecule hydrogen sulfide (H_2S) and PDO mutations could lead to the autosomal-recessive disorder ethylmalonic encephalopathy (EE). The proposed chemical mechanism for the PDO-catalyzed conversion of GSSH to GSH and sulfite is shown in Figure 1.13(A); Figure 1.13(B) shows the mechanism proposed for the catalytic mechanism-based PDO inhibition by γ-glutamyl-homocysteinyl-glycine (GSHcySH).[17] As shown, a simple replacement of CH_2 for S converts the substrate GSSH to the inhibitor GSHcySH, since the S–S bond is cleavable whereas the S–C bond is not.

1.2.2.4 A Catalytic Mechanism-based Inhibitor for the Glutathione Reductase-catalyzed Reaction

Glutathione reductase (GR) is a FAD-containing protein and catalyzes the NADPH-dependent reduction of glutathione disulfide (GSSG) to GSH, as shown in Figure 1.14(A).[18] It should be noted that FAD and its reduced forms present in GR mediate the one- or two-electron transfer between the end-electron donor NADPH and the end-electron acceptor $NADP^+$, and Figure 1.14(A) and (B) do not show the intervening electron transfer between NADPH and $NADP^+$. The GR-catalyzed reaction plays an important housekeeping role in the cellular redox homeostasis. Figure 1.14(B) depicts how a compound known as fluoro-M5 can irreversibly inhibit GR in a catalytic mechanism-based manner.[18]

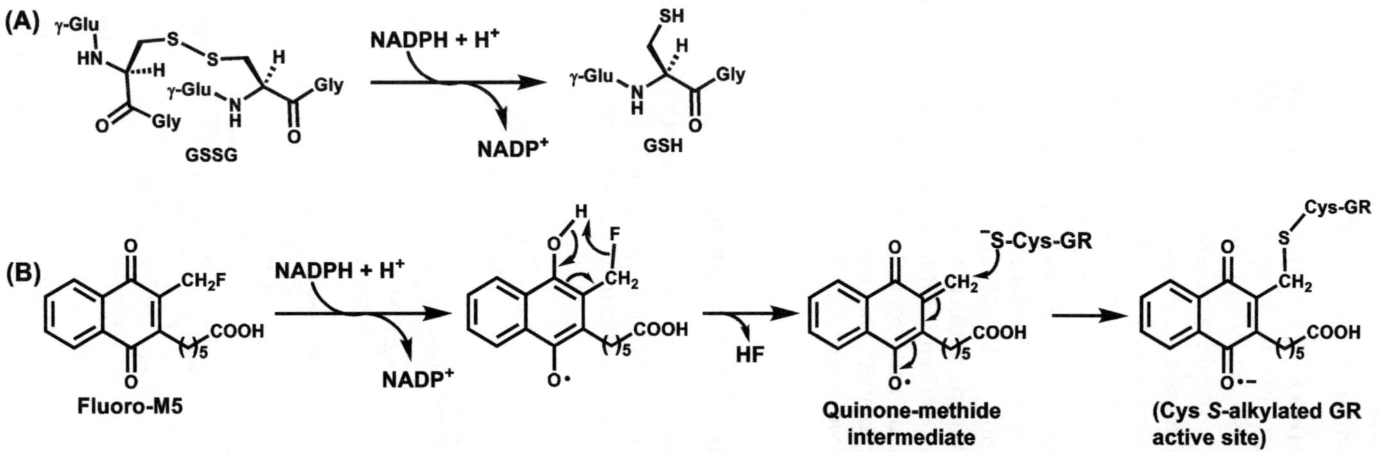

Figure 1.14 (A) The net reaction scheme of the GR-catalyzed NADPH-dependent reduction of GSSG to GSH. (B) The proposed mechanism for the GR-catalyzed conversion of fluoro-M5 to an electrophilic quinone-methide intermediate and the subsequent covalent interaction with an active site cysteine (Cys) residue side chain thiolate (S^-), leading to an irreversible GR inhibition.

1.2.2.5 A Catalytic Mechanism-based Inhibitor for the Proline Dehydrogenase-catalyzed Reaction

Proline dehydrogenase (PRODH) catalyzes the FAD-dependent oxidation (*via* a 2-electron path) of L-proline to Δ^1-pyrroline-5-carboxylate (Figure 1.15(A)), which constitutes the first step of L-proline catabolism.[9] The inhibitors of the PRODH-catalyzed reaction could be possible anti-cancer therapeutic agents and could also be employed as chemical probes for an enhanced understanding of the role of the PRODH-catalyzed reaction in cancer. Figure 1.15(B) depicts how an analog of L-proline known as L-T2C (L-thiazolidine-2-carboxylate) can irreversibly inhibit PRODH in a catalytic mechanism-based manner.[19] The chemical structure of L-T2C is fairly close to that of L-proline, however, due to the replacement of sulfur (S) for a ring methylene (CH_2) in the side chain of L-proline, the oxidation at the C5 atom becomes favored, thus leading to the formation of a covalent adduct between $FADH_2$ and the L-T2C-derived sulfonium intermediate and an inhibition of the PRODH-catalyzed reaction. It should be noted that this example of inhibition is analogous to the inactivation of LSD1-catalyzed reaction described above; in both cases the two-electron reduced coenzyme FAD (*i.e.* $FADH_2$) is covalently modified at the N5 position.

1.2.3 Applications with Hydrolases

Hydrolases catalyze the hydrolytic reactions on both protein and non-protein substrates, which are involved in a variety of important life processes. The substrates can be molecules harboring scissile amide (including β-lactam),

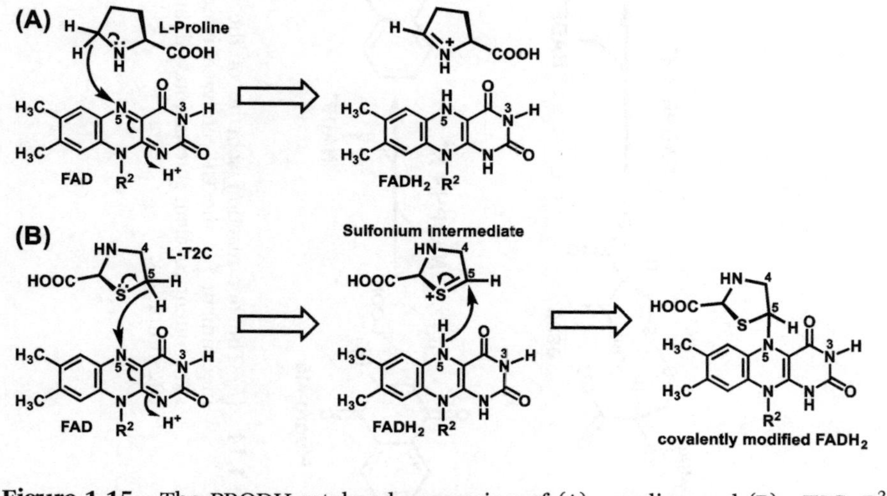

Figure 1.15 The PRODH-catalyzed processing of (A) L-proline and (B) L-T2C. R^2, ribosyl adenine dinucleotide. Please see Figure 5.11 in this book for the full chemical structures of FAD and $FADH_2$.

acetal, ester (including phosphate, sulfate, and thioester) or thioether linkages. The catalytic mechanism-based inhibitors have been developed for the hydrolase-catalyzed reactions on the amide, β-lactam, acetal, ester, and thioether substrates, as illustrated herein with notable examples. Similarly to the scenarios with the catalytic mechanism-based inhibitors described above, a time-dependent covalent (yet not non-covalent) irreversible inhibition has been observed with the catalytic mechanism-based inhibitors identified for the hydrolase-catalyzed reactions.

1.2.3.1 A Catalytic Mechanism-based Inhibitor for the Rhomboid-catalyzed Amide Hydrolysis Reaction

Rhomboids refer to a family of intra-membrane serine proteases that have been found to play an important role in multiple important cellular processes such as inter-cellular signaling and mitochondrial morphology.[20] The inhibitors for the rhomboid-catalyzed reaction can be employed to help clarify the functional roles of this enzymatic hydrolytic reaction and to be developed into potential therapeutic agents for human diseases. Figure 1.16(A) depicts the proposed chemical mechanism for a rhomboid-catalyzed peptide bond hydrolysis reaction, with the catalytically important active site serine (Ser) and histidine (His) residues indicated.[20] Figure 1.16(B) depicts the proposed *Escherichia coli* rhomboid GlpG-catalyzed processing of an isocoumarin-based catalytic mechanism-based inhibitor: following the initial covalent attack of the activated Ser side chain hydroxyl (OH) onto the lactone moiety of isocoumarin, the liberated benzylic chloride, which is also α to the methyl carboxylate, was susceptible to the nucleophilic attack by the His side chain imidazole nitrogen, furnishing the Ser/His doubly covalently modified catalytic dyad at the GlpG active site and the ensuing irreversible inhibition of the GlpG-catalyzed reaction.[20]

1.2.3.2 A Catalytic Mechanism-based Inhibitor for the β-Lactamase-catalyzed β-Lactam Hydrolysis Reaction

The active site Ser and His cross-linking by an isocoumarin-based catalytic mechanism-based inhibitor was also suggested to be the mode of the catalytic mechanism-based inhibition of *O*-aryloxycarbonyl hydroxamates against the reaction catalyzed by serine β-lactamases. While Figure 1.17(A) depicts the proposed chemical mechanism for the serine β-lactamase-catalyzed hydrolytic breakdown of a β-lactam substrate with the active site catalytic dyad residues serine (Ser) and histidine (His) indicated, Figure 1.17(B) depicts how serine β-lactamase can catalytically process one *O*-aryloxycarbonyl hydroxamate.[21] Following the initial covalent attack of the activated Ser side chain OH onto the aryloxycarbonyl carbon of the *O*-aryloxycarbonyl hydroxamate inhibitor, the resulting *O*-alkyloxycarbonyl hydroxamate covalent enzyme intermediate is susceptible to the depicted

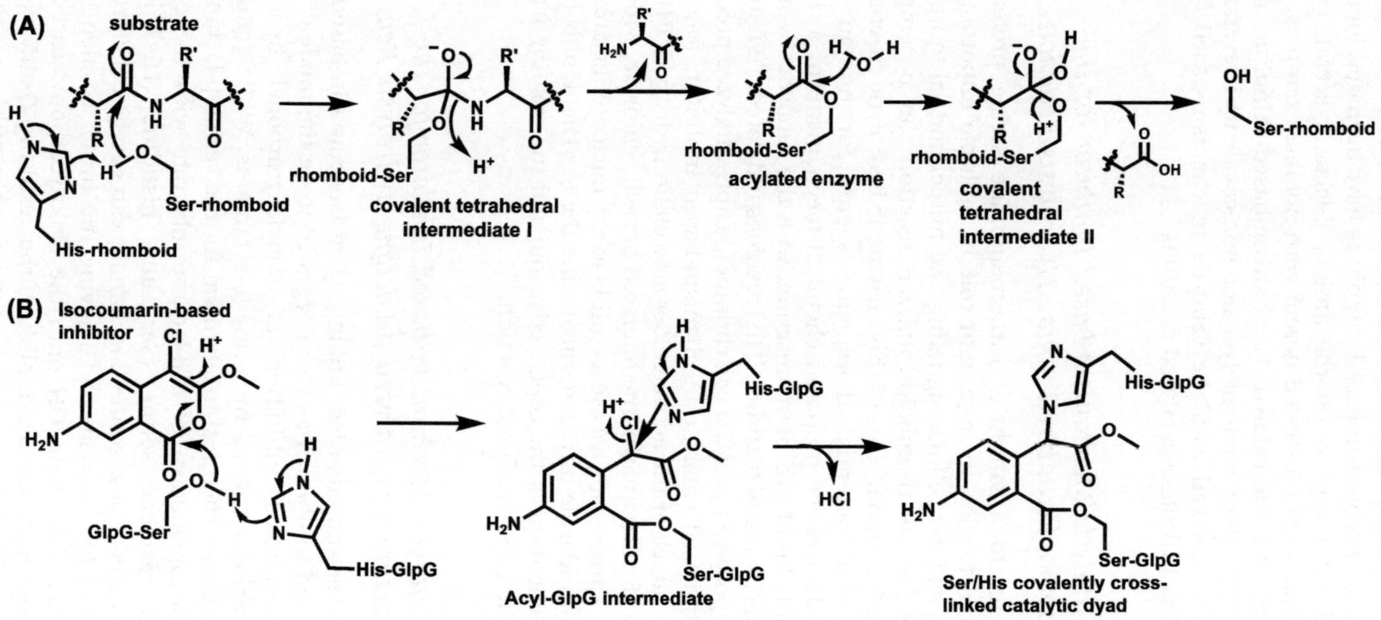

Figure 1.16 (A) The proposed chemical mechanism for a rhomboid-catalyzed peptide bond hydrolysis reaction. Serine (Ser) and histidine (His) refer to the active site catalytic dyad residues. R and R′ refer to the side chains flanking the scissile peptide bond. (B) The proposed *Escherichia coli* rhomboid GlpG-catalyzed processing of the depicted isocoumarin-based inhibitor, leading to the Ser/His covalently cross-linked catalytic dyad.

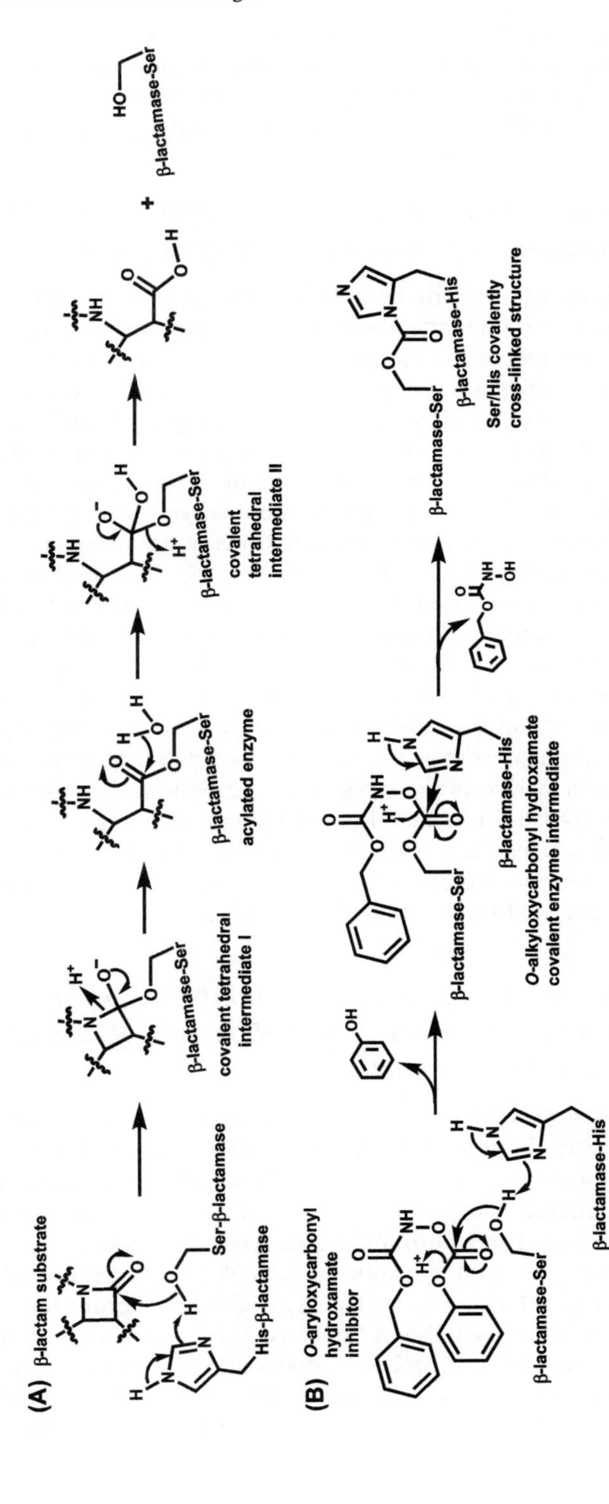

Figure 1.17 (A) The proposed chemical mechanism for the serine β-lactamase-catalyzed hydrolytic breakdown of a β-lactam substrate. Serine (Ser) and histidine (His) refer to the active site catalytic dyad residues. (B) The proposed serine β-lactamase-catalyzed processing of the depicted *O*-aryloxycarbonyl hydroxamate inhibitor, leading to the Ser/His covalently cross-linked catalytic dyad.

nucleophilic attack by the His side chain imidazole nitrogen onto the alkyloxycarbonyl carbon, furnishing the depicted Ser/His doubly covalently modified cyclic structure at the enzyme active site and the ensuing irreversible inhibition of the serine β-lactamase-catalyzed reaction.

1.2.3.3 A Catalytic Mechanism-based Inhibitor for the Glycoside Hydrolase-catalyzed Acetal Hydrolysis Reaction

Glycoside hydrolases catalyze the hydrolysis of the glycosidic bond (an acetal linkage) in various carbohydrate molecules. By counteracting the glycosyl-transferase-catalyzed reactions, the glycosidase-catalyzed reactions also play important roles in various life processes by helping to modulate the structure and function of various carbohydrate molecules. Figure 1.18(A) depicts the proposed chemical mechanism of the galactosidation reaction catalyzed by a retaining α-galactosidase.[22,23] The first stage of this double displacement mechanism involves the formation of the covalent enzyme intermediate with an aspartic acid residue *via* the depicted oxonium transition state, while the second stage involves the hydrolysis of this covalent enzyme intermediate liberating the glycone part of the substrate. The net result of this double displacement mechanism would be a retention of the α-configuration at the anomeric carbon. Figure 1.18(B) depicts the catalytic mechanism-based inhibition of this galactosidation reaction by the depicted cyclopropyl carbasugar.[22,23] A general acid-promoted cleavage of the C1–O bond leaves behind a cyclopropyl methyl cation, which would also exist as the depicted non-classical bicyclobutonium ion due to the participation of the strained C–C σ-bond to the positive charge delocalization at transition state. The subsequent reaction of this cationic structure with an aspartic acid side chain would generate a stable covalent enzyme adduct with one possible structure shown, leading to an irreversible inhibition of the enzymatic galactosidation reaction.

1.2.3.4 A Catalytic Mechanism-based Inhibitor for the Calcium-independent Phospholipase $A_2\beta$-Catalyzed Ester Hydrolysis Reaction

Phospholipase A_2's (PLA_2's) are one type of phospholipases that catalyze the hydrolysis of the ester linkage at the *sn*-2 position of glycerophospholipids to furnish free fatty acids and lysophospholipids, which play important role in regulating lipid metabolism and signaling. Calcium-independent PLA_2's ($iPLA_2$'s) are one of the three families of eukaryotic PLA_2's with the other two families being secretory and cytosolic PLA_2's. There are nine $iPLA_2$ family members including $iPLA_2\beta$. Figure 1.19(A) depicts the proposed chemical mechanism for the $iPLA_2\beta$-catalyzed ester hydrolysis reaction; Figure 1.19(B) depicts the proposed $iPLA_2\beta$-catalyzed processing of the depicted catalytic mechanism-based inhibitor known as (*S*)-BEL (the *S*-enantiomer of the racemic BEL).[24] As shown, following the initial acylation of the catalytic

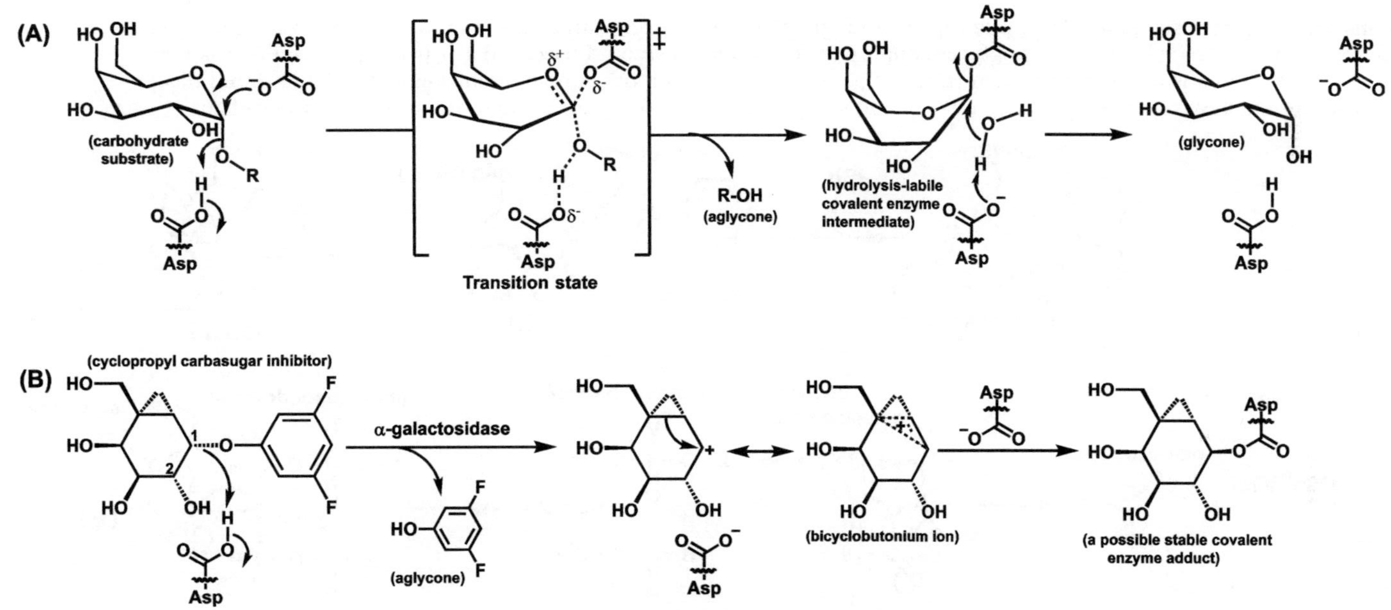

Figure 1.18 (A) The proposed chemical mechanism of the galactosidation reaction catalyzed by a retaining α-galactosidase. (B) The catalytic mechanism-based inhibition of the galactosidation reaction by the depicted cyclopropyl carbasugar inhibitor. Only one possible structure of the stable covalent enzyme adduct is shown.

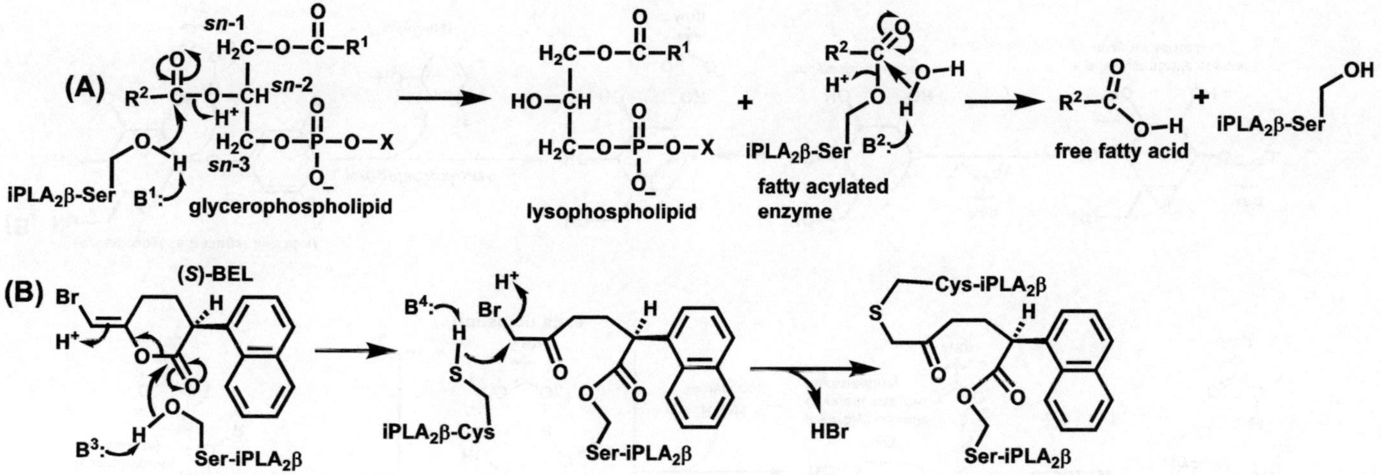

Figure 1.19 (A) The proposed chemical mechanism for the iPLA$_2$β-catalyzed reaction. R^1 and R^2, two fatty chains; X, a part of the polar head group of a glycerophospholipid substrate molecule; B^1: and B^2:, general bases at the active site of iPLA$_2$β; Ser, the catalytic serine residue. (B) The proposed processing of the depicted inhibitor (*S*)-BEL. In addition to the catalytic Ser, the active site cysteine (Cys) residue is also involved in this enzymatic process. B^3: and B^4:, general bases at the active site of iPLA$_2$β.

serine residue, the bromide derived from BEL was found to react intramolecularly with the active site cysteine residue. As a result, iPLA$_2$β is irreversibly inhibited due to the formation of a covalent cross-link between the active site serine and cysteine side chains.

It is worth noting that this mode of inhibition is reminiscent of that described in Sections 1.2.3.1 and 1.2.3.2 for the catalytic mechanism-based inhibitors respectively against the rhomboid- and the β-lactamase-catalyzed hydrolysis reactions.

1.2.3.5 *A Catalytic Mechanism-based Inhibitor for the* S-*Adenosylhomocysteine Hydrolase-catalyzed Thioether Hydrolysis Reaction*

S-Adenosylhomocysteine hydrolase (AdoHcyase) catalyzes the hydrolysis of the product of the widespread and important SAM-dependent enzymatic methylation reactions, *i.e.* *S*-adenosylhomocysteine (AdoHcy), to homocysteine and adenosine. As such, the AdoHcyase-catalyzed reaction plays an important role in regulating cellular (patho)physiological processes and represents a potential therapeutic target for human diseases.[25,26]

Figure 1.20 depicts the proposed chemical mechanism of the AdoHcyase-catalyzed hydrolysis reaction.[25] As shown, the key step is the proton abstraction at the C4′ position, which would induce the splitting of the AdoHcy skeleton with the production of homocysteine. Additionally, while the first step involves the reduction of NAD$^+$ to NADH, the last step involves the reoxidation of NADH back to NAD$^+$.

Figure 1.21 depicts the AdoHcyase-catalyzed processing of the catalytic mechanism-based inhibitor fluoroneplanocin A.[26] It is worth noting that, besides causing the covalent labeling of AdoHcyase, leading to an irreversible inhibition, the action of this inhibitor would also cause a reversible inhibition *via* NAD$^+$ depletion, which could be reversed by NAD$^+$ replenishment.

1.2.4 Applications with Lyases

Lyases catalyze the group elimination reactions on substrate molecules. However, in this section, the development of the catalytic mechanism-based inhibitors for several lyation reactions (*i.e.* group elimination reactions) catalyzed not just by lyases, but also by other enzymes (*e.g.* decarboxylase), are covered, illustrated with notable examples. These enzymatic reactions play important regulatory roles in a variety of important life processes. As for the scenarios with the above-described catalytic mechanism-based inhibitors, a time-dependent covalent (and also non-covalent) irreversible inhibition has been observed with the catalytic mechanism-based inhibitors identified for the enzymatic lyation reactions.

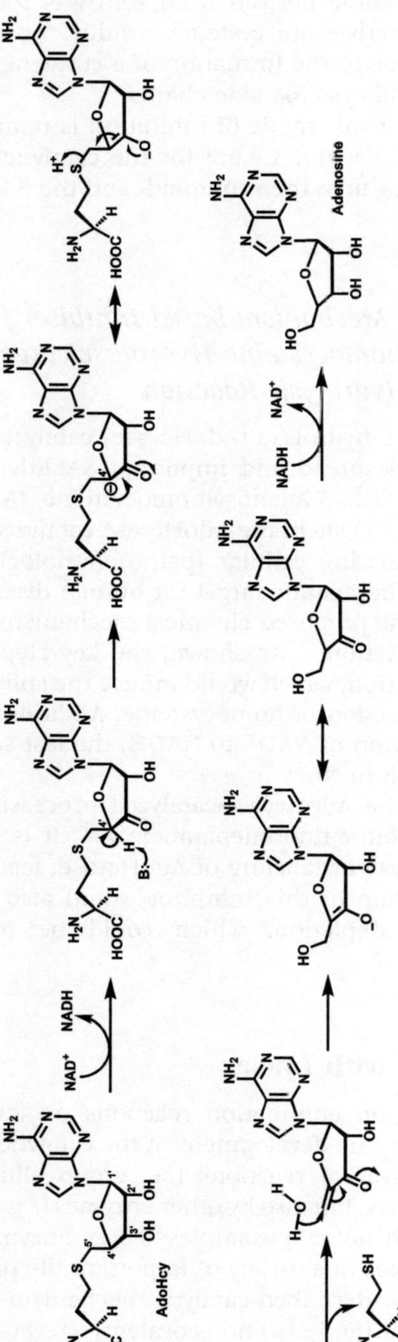

Figure 1.20 The proposed chemical mechanism of the AdoHcyase-catalyzed hydrolysis reaction. B:, a general base at the active site of AdoHcyase. Please see other figures (*e.g.* Figure 3.7) in this book for the full chemical structures of NAD^+ and NADH.

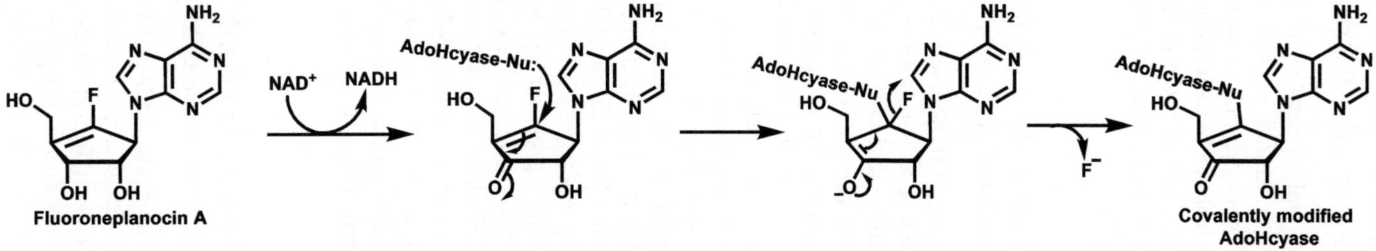

Figure 1.21 The AdoHcyase-catalyzed processing of the catalytic mechanism-based inhibitor fluoroneplanocin A. Nu:, a nucleophile from the active site of AdoHcyase. Please see other figures (*e.g.* Figure 3.7) in this book for the full chemical structures of NAD^+ and NADH.

1.2.4.1 A Catalytic Mechanism-based Inhibitor for the Isocitrate Lyase-catalyzed Retro-aldol Reaction

Isocitrate lyase catalyzes the retro-aldol cleavage of isocitrate to form glyoxylate and succinate, which is the first enzymatic reaction for the pathophysiologically important glyoxylate shunt that is present in non-mammalian organisms.[27] Figure 1.22(A) depicts the chemical mechanism proposed for the isocitrate lyase-catalyzed reaction.[27] Figure 1.22(B) depicts the proposed chemical mechanism for the isocitrate lyase-catalyzed processing of 2-vinyl D-isocitrate leading to irreversible enzyme inhibition.[27]

1.2.4.2 A Catalytic Mechanism-based Inhibitor for the Aminomutase-catalyzed Reaction

A 4-methylideneimidazole-5-one (MIO)-based aminomutase catalyzes the reversible inter-conversion between a proteinogenic aromatic α-amino acid and its corresponding β-amino acid, which plays an important role in regulating the biosynthesis of the important family of β-amino acids.[28] Figure 1.23(A) depicts the chemical mechanism of the enzymatic reaction specifically catalyzed by the tyrosine aminomutase (TAM).[28] The key steps of this reversible enzymatic transformation are the depicted β-deprotonation for the forward reaction and the α-deprotonation for the backward reaction (not depicted) on α-tyrosine and β-tyrosine, respectively, with the corresponding cleavage of the α- and β-amino groups in the form of the covalent adduct with cofactor MIO. Figure 1.23(B) depicts the TAM-catalyzed processing of an analog of β-tyrosine (*i.e.* α-difluoro-β-Tyr) with the formation of its stable covalent adduct with cofactor MIO through its β-amino group.[28] The stability of this intermediate presumably results from the lack of an α-proton for deprotonation. This intermediate stalled at TAM active site could be regarded as a tight-binding bi-substrate analog inhibitor with its two structural parts derived from α-difluoro-β-Tyr and cofactor MIO, consistent with the experimentally observed irreversible inhibition of α-difluoro-β-Tyr against the TAM-catalyzed reaction. Of note, α-difluoro-β-Tyr is another example of a catalytic mechanism-based inhibitor exerting a non-covalent inhibition.

1.2.4.3 A Catalytic Mechanism-based Inhibitor for the Malonate Semialdehyde Decarboxylase-catalyzed Reaction

Malonate semialdehyde decarboxylase (MSAD) catalyzes the decarboxylation of malonate semialdehyde to form acetaldehyde. MSAD also possesses a hydratase activity, which has been found to be essential for the catalytic mechanism-based inhibition of 3-chloro-propargylate. Figure 1.24(A) depicts the proposed chemical mechanism of the MSAD-catalyzed decarboxylation of malonate semialdehyde with the formation of acetaldehyde and carbon

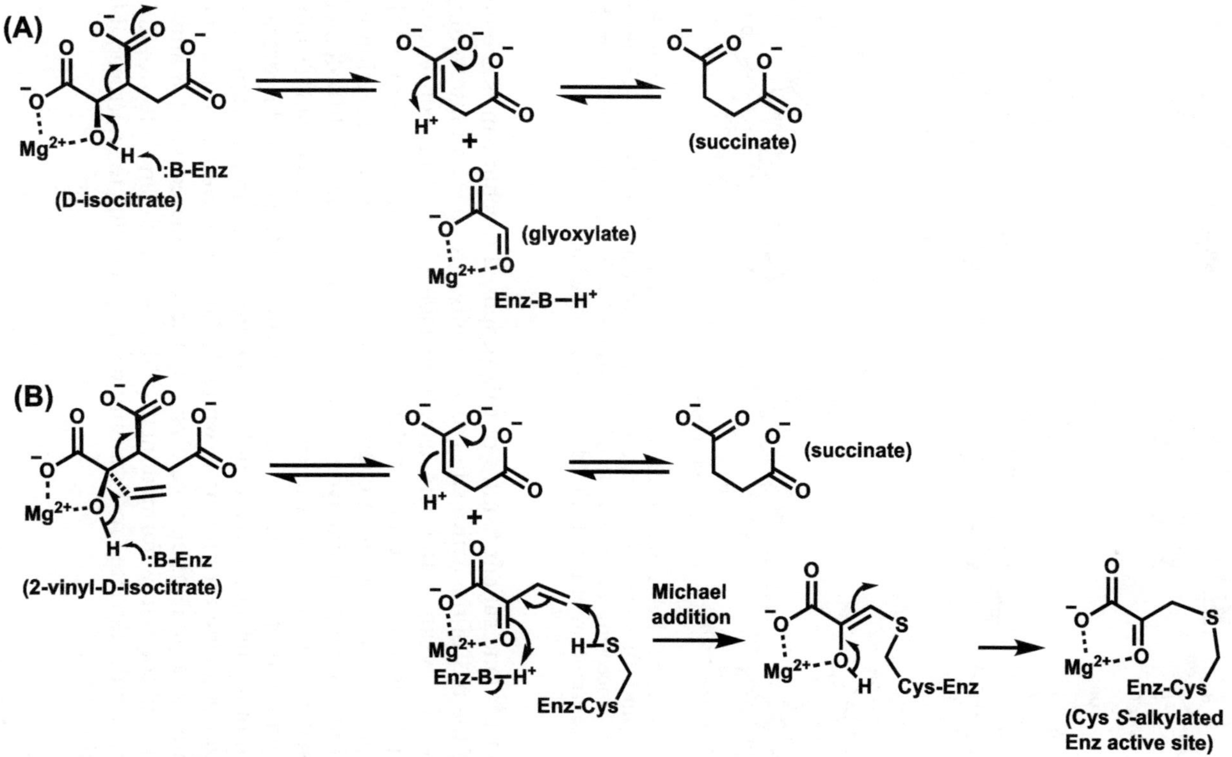

Figure 1.22 (A) The proposed chemical mechanism for the isocitrate lyase-catalyzed retro-aldol reaction. Enz, enzyme, *i.e.* isocitrate lyase; B:, a general base at the active site of isocitrate lyase. (B) The chemical mechanism proposed for the isocitrate lyase-catalyzed processing of 2-vinyl D-isocitrate leading to the covalent alkylation of an active site cysteine (Cys) residue and the ensuing irreversible enzyme inhibition. Enz and B:, defined as in (A).

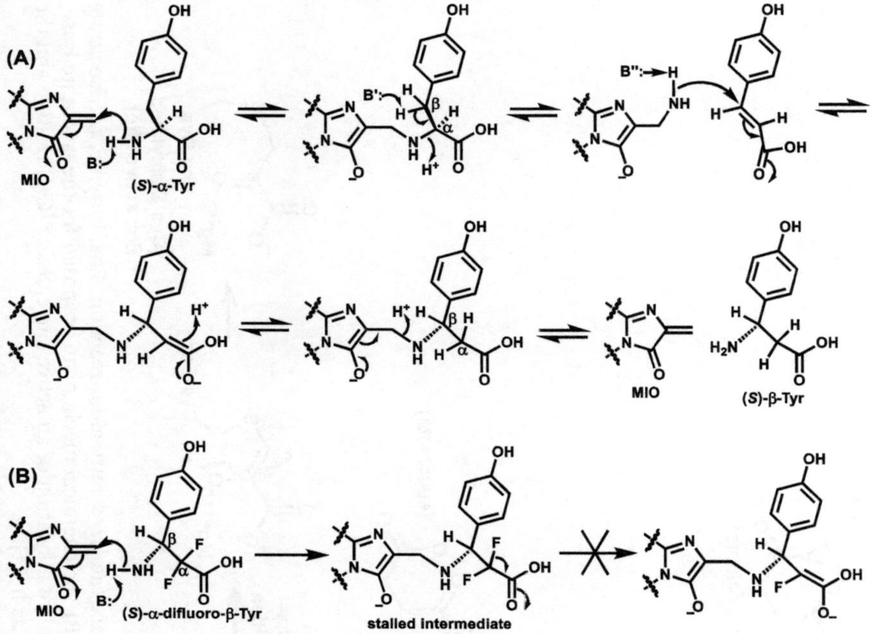

Figure 1.23 (A) The chemical mechanism of the reaction catalyzed by TAM. B:, B′:, and B″:, general bases at the active site of TAM. (B) The TAM-catalyzed processing of an analog of β-tyrosine (*i.e.* α-difluoro-β-Tyr) with the formation of the depicted stalled intermediate.

dioxide (CO_2).[29] One salient feature of this proposed mechanism is the depression of the pK_b of the active site Pro-1 α-amino group *via* an intervening water molecule by an active site aspartate side chain. The consequent enhanced proportion of the protonated form of this amino group would contribute more to the negative charge stabilization along the reaction coordinate of malonate semialdehyde decarboxylation. Figure 1.24(B) depicts the proposed mechanism for the catalytic mechanism-based MSAD inhibition by 3-chloro-propargylate.[29] As depicted, the MSAD-catalyzed hydration (or addition of an activated water molecule) onto this inhibitory compound gives rise to the allene intermediate, which is further converted to the enol intermediate *via* a general acid catalysis by the protonated Pro-1 α-amino group. The subsequently formed acid chloride or ketene electrophilic intermediate would alkylate Pro-1's α-amino group, leading to an irreversible MSAD inhibition.

1.2.5 Applications with Isomerases

Isomerases catalyze a variety of reactions involved in important life processes. In this section, the development of the catalytic mechanism-based

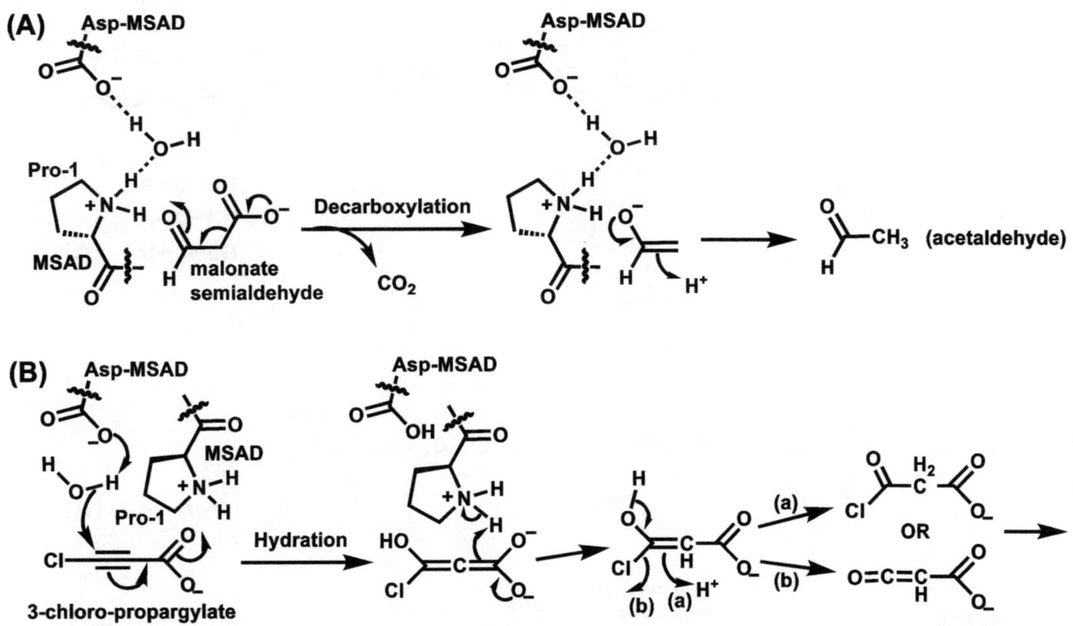

Figure 1.24 (A) The proposed chemical mechanism of the MSAD-catalyzed decarboxylation of malonate semialdehyde with the formation of acetaldehyde and CO_2. (B) The proposed mechanism for the hydration-initiated catalytic mechanism-based MSAD inhibition by 3-chloro-propargylate. Pro-1, the first residue of MSAD.

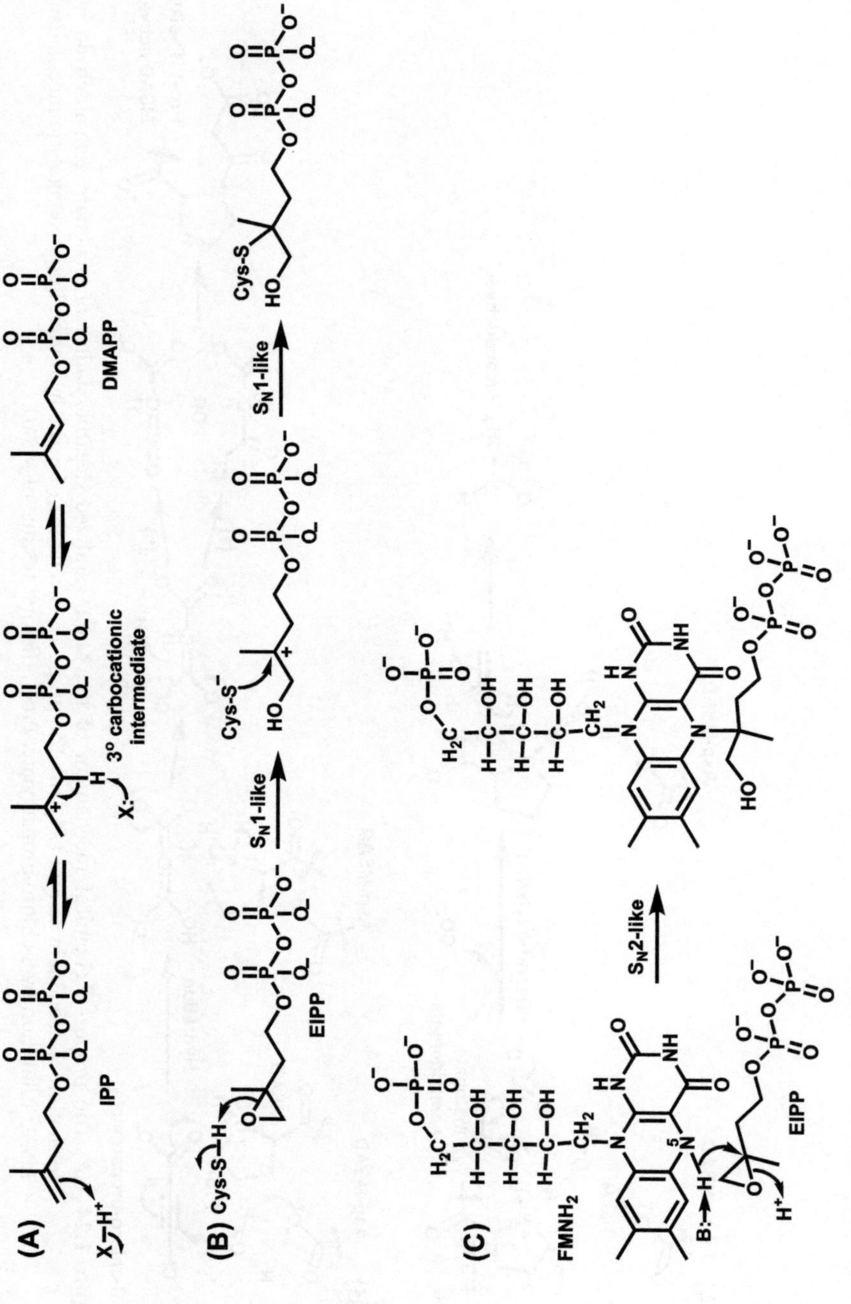

(A)

X–H⁺ IPP ⇌ X: 3° carbocationic intermediate ⇌ DMAPP

(B) Cys-S–H

EIPP → S_N1-like → Cys-S⁻ HO → S_N1-like → Cys-S HO

(C)

FMNH$_2$

H$_2$C O–P–O⁻ H–C–OH H–C–OH H–C–OH CH$_2$ N NH O 5 N B: H O H⁺ EIPP

S_N2-like

H$_2$C O–P–O⁻ H–C–OH H–C–OH H–C–OH CH$_2$ N NH O HO

inhibitors for the isomerase-catalyzed reaction and for the reaction catalyzed by a racemase is covered, illustrated with notable examples. As in the scenarios with the above-described catalytic mechanism-based inhibitors, a time-dependent covalent (and also non-covalent) irreversible inhibition has been observed with the catalytic mechanism-based inhibitors identified for the enzymatic isomerization and racemization reactions.

1.2.5.1 A Catalytic Mechanism-based Inhibitor for the Isopentenyl Diphosphate Isomerase-catalyzed Reaction

Isopentenyl diphosphate isomerase (IDI) can be present as the type-1 (IDI-1) or the type-2 (IDI-2) isoform in different organisms of the three evolutionary life forms. Even though IDI-1 and IDI-2 show no sequence homology, they both catalyze the same reaction, *i.e.* the inter-conversion between isopentenyl diphosphate (IPP) and dimethylallyl diphosphate (DMAPP), the two fundamental building blocks in isoprenoid biosynthesis. It should be noted that, while IDI-1 only requires a divalent metal cation (*e.g.* Mg^{2+}) for activity, IDI-2 is a flavoprotein harboring not only Mg^{2+} but also flavin mononucleotide (FMN) and NADPH; despite this, the IDI-1- and IDI-2-catalyzed isomerization reactions entail the same chemical mechanism: protonation–deprotonation with a tertiary (3°) carbocationic intermediate, as depicted in Figure 1.25(A).[30,31]

Figure 1.25(B) depicts the IDI-1-catalyzed processing of the catalytic mechanism-based inhibitor 3,4-epoxy-3-methylbutyl diphosphate (EIPP) (racemic). As depicted, the oxirane ring opening in EIPP gives rise to a tertiary carbocation whose reaction with an active site cysteine (Cys) side chain thiol (SH) would lead to a covalent inhibition of the IDI-1-catalyzed reaction. However, as depicted in Figure 1.25(C) for the IDI-2-catalyzed processing of the racemic EIPP, the oxirane ring opening in EIPP now follows a S_N2-like mode, *i.e.* nucleophilic substitution between N5 of $FMNH_2$ and the indicated oxirane ring carbon. The result would be the formation of the depicted covalent adduct which would bind to IDI-2 active site non-covalently. Since EIPP was found to also inhibit the IDI-2-catalyzed reaction in a time-dependent manner, this covalent adduct would bind tightly to the IDI-2 active site, presumably behaving as a bi-substrate analog inhibitor.[31]

Figure 1.25 (A) The proposed chemical mechanism for the IDI-catalyzed isomerization reaction between IPP and DMAPP *via* the depicted carbocationic intermediate. $X–H^+$ could be a protonated active site residue side chain (*e.g.* Cys-SH) or the protonated N5 of the reduced FMN (*i.e.* $FMNH_2$). Mg^{2+} binds to the negatively charged diphosphate part. (B) The proposed S_N1-like mechanism for the IDI-1-catalyzed processing of the racemic EIPP. (C) The proposed S_N2-like mechanism for the IDI-2-catalyzed processing of the racemic EIPP. B:, a general base at the IDI-2 active site.

1.2.5.2 A Catalytic Mechanism-based Inhibitor for the Alanine Racemase-catalyzed Reaction

The bacterial enzyme alanine racemase catalyzes the PLP-dependent interconversion between L-alanine and D-alanine, which plays an essential role in

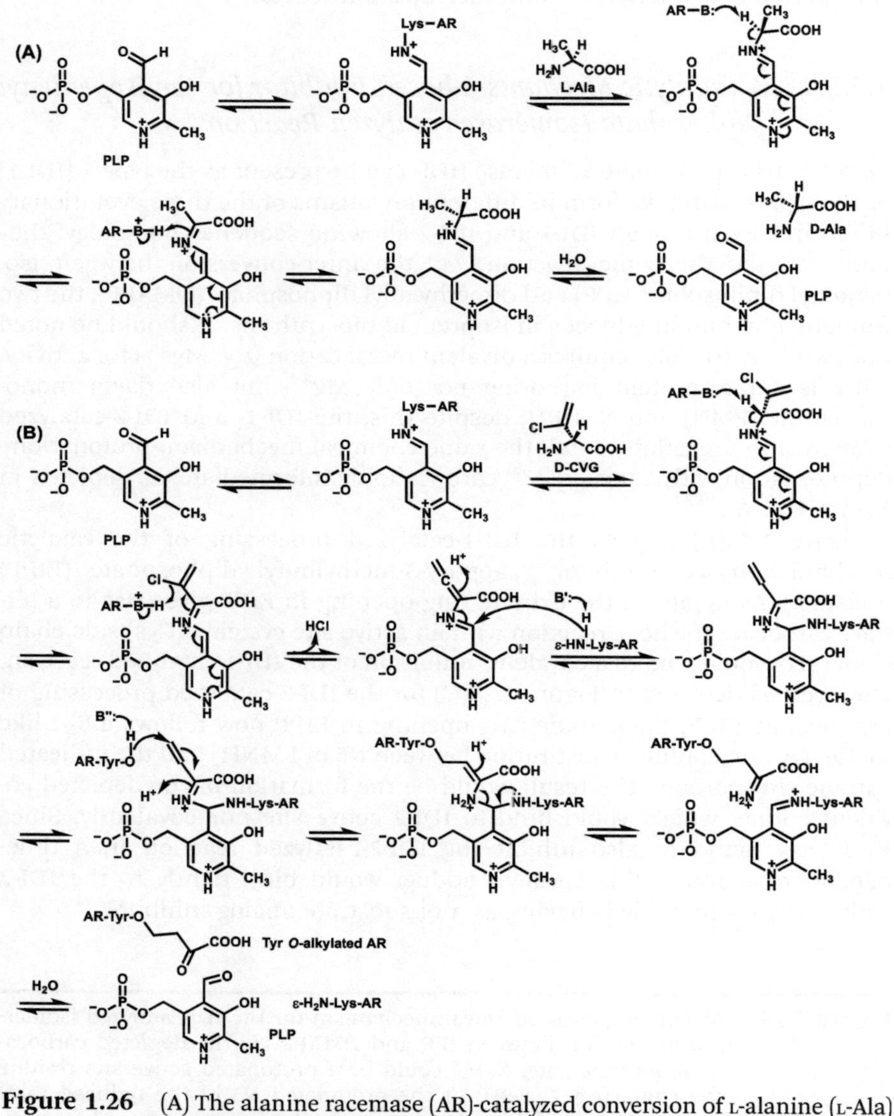

Figure 1.26 (A) The alanine racemase (AR)-catalyzed conversion of L-alanine (L-Ala) to D-alanine (D-Ala). B:, a general base at the AR active site. (B) The AR-catalyzed processing of D-chlorovinylglycine (D-CVG) leading to covalently modified AR and the consequent irreversible inhibition of the AR-catalyzed reaction. B:, B′:, and B″:, general bases at the AR active site; Tyr, tyrosine; Lys, lysine.

the bacterial cell wall peptidoglycan biosynthesis by furnishing the essential building block D-alanine. Therefore, the inhibitors of this enzymatic reaction are potential antibacterials. Figure 1.26(A) depicts the alanine racemase-catalyzed conversion of L-alanine to D-alanine.[32] The key mechanistic step is the facial-specific protonation at the α-carbon furnishing the D-alanine. Figure 1.26(B) depicts the alanine racemase-catalyzed processing of D-chlorovinylglycine (an analog of D-alanine), leading to covalently modified alanine racemase and the consequent irreversible inhibition of the alanine racemase-catalyzed reaction.[32] The key mechanistic steps include the formation of the electrophilic allenic intermediate following the elimination of Cl^- in the form of HCl; and the conjugate addition of an active site tyrosine side chain OH onto the terminal carbon of the conjugate system derived from the rearrangement of the allenic intermediate.

1.2.6 Applications with Ligases

Ligases catalyze a variety of ligation reactions involved in important life processes; however, catalytic mechanism-based inhibitors have not yet been developed for ligase-catalyzed reactions.

References

1. J. A. Doudna and J. R. Lorsch, *Nat. Struct. Mol. Biol.*, 2005, **12**, 395.
2. K. A. Johnson, *Biochim. Biophys. Acta*, 2010, **1804**, 1041.
3. J. Zhong and W. C. Groutas, *Curr. Top. Med. Chem.*, 2004, **4**, 1203.
4. J. G. Voet, D. Voet and C. W. Pratt, *Fundamentals of biochemistry: life at the molecular level*, Wiley, Hoboken, 5th edn, 2016.
5. A. A. Sauve and D. Y. Youn, *Curr. Opin. Chem. Biol.*, 2012, **16**, 535.
6. W. Zheng, *Plant Sci.*, 2020, **293**, 110434.
7. Y. Jiang, J. Liu, D. Chen, L. Yan and W. Zheng, *Trends Pharmacol. Sci.*, 2017, **38**, 459.
8. Y. Wang, Y. M. E. Fung, W. Zhang, B. He, M. W. H. Chung, J. Jin, J. Hu, H. Lin and Q. Hao, *Cell Chem. Biol.*, 2017, **24**, 339.
9. H. Lee, E. H. Doud, R. Wu, R. Sanishvili, J. I. Juncosa, D. Liu, N. L. Kelleher and R. B. Silverman, *J. Am. Chem. Soc.*, 2015, **137**, 2628.
10. Y. Yao, P. Chen, J. Diao, G. Cheng, L. Deng, J. L. Anglin, B. V. V. Prasad and Y. Song, *J. Am. Chem. Soc.*, 2011, **133**, 16746.
11. N. Allocati, M. Masulli, C. D. Ilio and L. Federici, *Oncogenesis*, 2018, 7, 8.
12. K. Sato, Y. Kunitomo, Y. Kasai, S. Utsumi, I. Suetake, S. Tajima, S. Ichikawa and A. Matsuda, *ChemBioChem*, 2018, **19**, 865.
13. J. C. Culhane and P. A. Cole, *Curr. Opin. Chem. Biol.*, 2007, **11**, 561.
14. L. M. Szewczuk, J. C. Culhane, M. Yang, A. Majumdar, H. Yu and P. A. Cole, *Biochemistry*, 2007, **46**, 6892.
15. G. Dong, L.-R. Lin, L.-Y. Xu and E.-M. Li, *J. Inorg. Biochem.*, 2020, **211**, 111204.

16. A. D. Findlay, J. S. Foot, A. Buson, M. Deodhar, A. G. Jarnicki, P. M. Hansbro, G. Liu, H. Schilter, C. I. Turner, W. Zhou and W. Jarolimek, *J. Med. Chem.*, 2019, **62**, 9874.
17. O. Kabil, N. Motl, M. Strack, J. Seravalli, N. Metzler-Nolte and X. R. Banerjee, *J. Biol. Chem.*, 2018, **293**, 12429.
18. H. Bauer, K. Fritz-Wolf, A. Winzer, S. Kühner, S. Little, V. Yardley, H. Vezin, B. Palfey, R. H. Schirmer and E. Davioud-Charvet, *J. Am. Chem. Soc.*, 2006, **128**, 10784.
19. A. C. Campbell, D. F. Becker, K. S. Gates and J. J. Tanner, *ACS Chem. Biol.*, 2020, **15**, 936.
20. K. R. Vinothkumar, K. Strisovsky, A. Andreeva, Y. Christova, S. Verhelst and M. Freeman, *EMBO J.*, 2010, **29**, 3797.
21. R. Tilvawala, M. Cammarata, S. A. Adediran, J. S. Brodbelt and R. F. Pratt, *Biochemistry*, 2015, **54**, 7375.
22. S. Chakladar, Y. Wang, T. Clark, L. Cheng, S. Ko, D. J. Vocadlo and A. J. Bennet, *Nat. Commun.*, 2014, **5**, 5590.
23. C. Adamson, R. J. Pengelly, S. S. K. Abadi, S. Chakladar, J. Draper, R. Britton, T. M. Gloster and A. J. Bennet, *Angew. Chem., Int. Ed.*, 2016, **55**, 14978.
24. C. M. Jenkins, J. Yang and R. W. Gross, *Biochemistry*, 2013, **52**, 4250.
25. Y. Kusakabe, M. Ishihara, T. Umeda, D. Kuroda, M. Nakanishi, Y. Kitade, H. Gouda, K. T. Nakamura and N. Tanaka, *Sci. Rep.*, 2015, **5**, 16641.
26. K. M. Lee, W. J. Choi, Y. Lee, H. J. Lee, L. X. Zhao, H. W. Lee, J. G. Park, H. O. Kim, K. Y. Hwang, Y.-S. Heo, S. Choi and L. S. Jeong, *J. Med. Chem.*, 2011, **54**, 930.
27. T. V. Pham, A. S. Murkin, M. M. Moynihan, L. Harris, P. C. Tyler, N. Shetty, J. C. Sacchettini, H.-L. Huang and T. D. Meek, *Proc. Natl. Acad. Sci. U. S. A.*, 2017, **114**, 7617.
28. C. V. Christianson, T. J. Montavon, G. M. Festin, H. A. Cooke, B. Shen and S. D. Bruner, *J. Am. Chem. Soc.*, 2007, **129**, 15744.
29. J. J. Almrud, G. J. Poelarends, W. H. Johnson Jr., H. Serrano, M. L. Hackert and C. P. Whitman, *Biochemistry*, 2005, **44**, 14818.
30. T. Nagai, H. Unno, M. W. Janczak, T. Yoshimura, C. D. Poulter and H. Hemmi, *Proc. Natl. Acad. Sci. U. S. A.*, 2011, **108**, 20461.
31. T. Hoshino, H. Tamegai, K. Kakinuma and T. Eguchi, *Bioorg. Med. Chem.*, 2006, **14**, 6555.
32. N. A. Thornberry, H. G. Bull, D. Taub, K. E. Wilson, G. Gimenqz-Gallego, A. Rosegay, D. D. Soderman and A. A. Patchett, *J. Biol. Chem.*, 1991, **266**, 21657.

CHAPTER 2

Catalytic Intermediate-based Design

2.1 Mode of Working

Many enzyme-catalyzed reactions proceed *via* one or more kinetically stable catalytic intermediates, for example, the acyl–enzyme covalent intermediate along the reaction coordinate of the serine protease-catalyzed hydrolytic reaction. The structural features of such catalytic intermediates have been employed as templates to design their close structural analogs so that such designed molecular entities would be able to compete with the native catalytic intermediates for binding to enzyme active sites, leading to inhibition of the corresponding enzymatic reaction.

A catalytic intermediate is the product of an earlier enzymatic step along the reaction coordinate and the substrate of the immediate downstream enzymatic step, with its structural features postulated to be closer to those of the more stable transition state of the two steps pursuant to the Hammond postulate.[1,2] Therefore, the intermediate-based design strategy has been quite successful with many enzymatic reactions, with notable examples presented in the following sections.

2.2 Applications

As for the topic organization of Chapter 1, notable examples of the catalytic intermediate-based inhibitors for each of the six types of the enzymatic reactions (*i.e.* the reactions catalyzed by transferases, oxidoreductases, hydrolases, lyases, isomerases, and ligases) are elaborated in this section.

2.2.1 Applications with Transferases

In this section, notable examples of transferase-catalyzed reactions have been chosen to illustrate how the catalytic intermediate-based inhibitors can

Active Site-directed Enzyme Inhibitors: Design Concepts
By Weiping Zheng

Published by the Royal Society of Chemistry, www.rsc.org

be developed for this class of enzymatic reactions. It should be noted that, unlike the scenario with catalytic mechanism-based inhibitors described above where either covalent or non-covalent inhibition can be achieved with a transferase-catalyzed reaction, only non-covalent inhibition would be theoretically possible and has indeed been observed with catalytic intermediate-based inhibitors for transferase-catalyzed reactions. Additionally, contingent upon the binding tightness of an inhibitor at the enzyme active site, either reversible or irreversible inhibition of the normal enzymatic catalysis would be observed.

2.2.1.1 A Catalytic Intermediate-based Inhibitor for the DNA (Cytosine C5)-methyltransferase (DNA C5-MTase)-catalyzed Methylation Reaction

DNA (cytosine C5)-methyltransferase (DNA C5-MTase) catalyzes the *S*-adenosyl-L-methionine (SAM)-dependent methylation at the C5 position of certain cytosine residues in a DNA strand, as shown in Figure 2.1(A).[3] A key mechanistic feature of this enzymatic reaction is the formation of the depicted 5-methyl-5,6-dihydrocytosine intermediate which is covalently attached at the C6 position onto the side chain thiolate (S^-) of an active site cysteine residue of DNA C5-MTase. By replacing the C5 carbon of cytosine with NH, the resulting analog (*i.e.* 5,6-dihydro-5-azacytosine, shown in Figure 2.1(B)) could be regarded as a mimic of the afore-mentioned catalytic intermediate when incorporated in a DNA molecule. Presumably owing to this structural mimicry, an oligodeoxyribonucleotide harboring this analog was found to form a tight yet reversible complex with DNA C5-MTase, leading to potent inhibition of the DNA C5-MTase-catalyzed methylation reaction.[3]

2.2.1.2 A Catalytic Intermediate-based Inhibitor for the Glycinamide Ribonucleotide Transformylase-catalyzed Formyl Transfer Reaction

Glycinamide ribonucleotide (GAR) transformylase (GAR Tfase) catalyzes the formylation of the primary amine of β-GAR with 10-formyl-tetrahydrofolate (10-formyl-THF) as the formyl donor leading to the formation of formyl β-GAR, which is an important step in the *de novo* purine biosynthetic pathway. The inhibition of the GAR Tfase-catalyzed formylation reaction has been regarded as one important strategy for cancer drug development. Figure 2.2(A) depicts the proposed chemical mechanism for the GAR Tfase-catalyzed formylation reaction.[4] One salient mechanistic feature is the formation of the depicted tetrahedral intermediate. Figure 2.2(B) shows the chemical structure of an analog of 10-formyl-THF, *i.e.* 10-formyl-5,8,10-trideazafolic acid (10-formyl-TDAF), which was found to be a potent inhibitor of the GAR Tfase-catalyzed formylation reaction. It was further found that

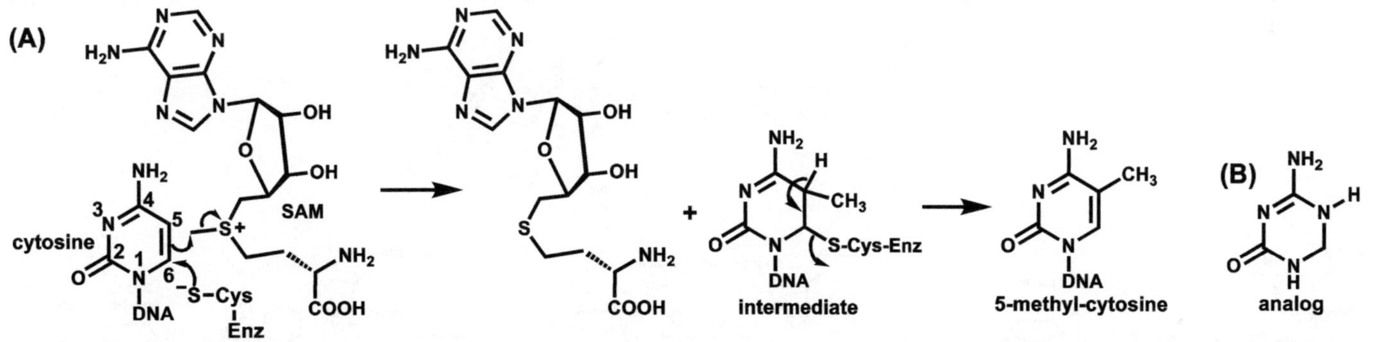

Figure 2.1 (A) The proposed chemical mechanism of the SAM-dependent methylation reaction catalyzed by DNA C5-MTase. Enz, enzyme (*i.e.* DNA C5-MTase); Cys, cysteine; "intermediate" refers to the 5-methyl-5,6-dihydrocytosine intermediate. (B) The chemical structure of the designed analog of cytosine which would mimic the 5-methyl-5,6-dihydrocytosine intermediate when incorporated into a DNA molecule.

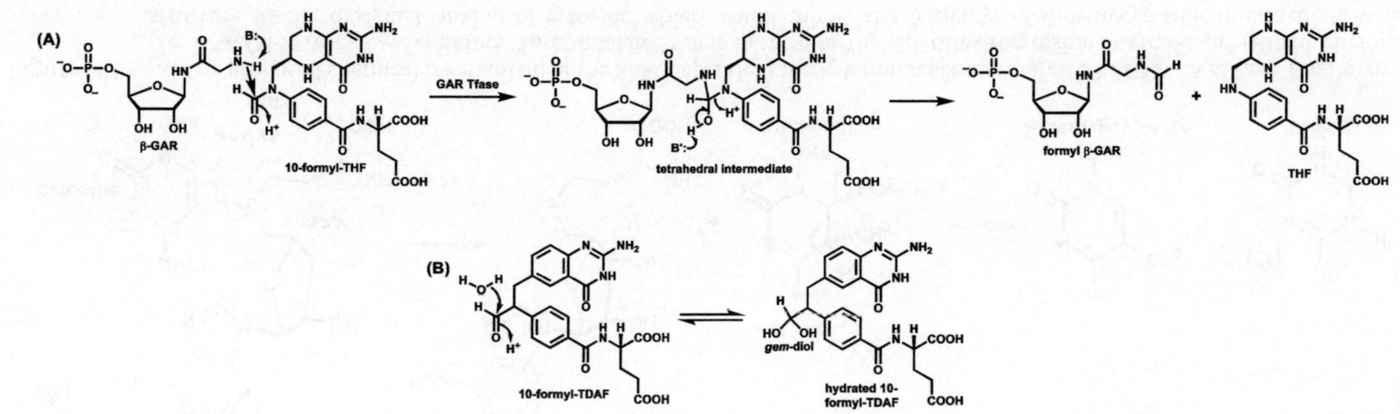

Figure 2.2 (A) The proposed chemical mechanism of the 10-formyl-THF-dependent formylation reaction catalyzed by GAR Tfase. B: and B′:, general bases at the active site of GAR Tfase. (B) The chemical structure of 10-formyl-TDAF, an analog of 10-formyl-THF, whose hydrated gem-diol form would structurally mimic the tetrahedral intermediate depicted in (A).

this compound would be present as its hydrated *gem*-diol form at the GAR Tfase active site, and as such it could serve as a structural mimic of the tetrahedral intermediate depicted in Figure 2.2(A).[4]

2.2.1.3 A Catalytic Intermediate-based Inhibitor for the γ-Glutamyl Transpeptidase-catalyzed Reaction

γ-Glutamyl transpeptidase (γGTase) catalyzes the transfer of a γ-glutamyl group from a γ-glutamyl-containing compound (*e.g.* glutathione) to a nucleophile (*e.g.* a dipeptide) and plays an important role in regulating important cellular processes such as glutathione metabolism. Figure 2.3(A) depicts the proposed chemical mechanism of the γGTase-catalyzed reaction.[5] As depicted, one salient mechanistic feature is the formation of two oxyanion tetrahedral intermediates along the reaction coordinate respectively leading to the formation of the γ-glutamylated active site hydroxyl and the γ-glutamylated nucleophile. Given this mechanism, a γ-boronic acid analog of L-glutamic acid known as γ-boroGlu (depicted in Figure 2.3(B)) was designed and was found to potently inhibit the γGTase-catalyzed reaction. The potent inhibitory action of γ-boroGlu would presumably result from the ability of its reaction product with the γGTase active site hydroxyl to structurally mimic the two oxyanion tetrahedral intermediates, as shown in Figure 2.3.[5]

2.2.2 Applications with Oxidoreductases

As mentioned in Chapter 1, oxidoreductases include oxidases, reductases, and dehydrogenases and catalyze a variety of oxidation or reduction reactions involved in various life processes. However, catalytic intermediate-based inhibitors have been developed only for the dehydrogenase-catalyzed reactions such as the examples detailed below. Unlike the scenario with the catalytic mechanism-based inhibitors where either covalent or non-covalent inhibition can be achieved with an oxidoreductase-catalyzed reaction, as described above, only non-covalent inhibition would be theoretically possible and has been observed with the catalytic intermediate-based inhibitors for the dehydrogenase-catalyzed reactions. For the tight binding inhibitors, functionally irreversible inhibition would also be observed.

2.2.2.1 A Catalytic Intermediate-based Inhibitor for the Inosine Monophosphate Dehydrogenase-catalyzed Reaction

Figure 2.4(A) depicts the proposed chemical mechanism for the inosine monophosphate dehydrogenase (IMPDH)-catalyzed nicotinamide adenine dinucleotide (NAD^+)-dependent dehydrogenation reaction.[6] As depicted, one salient mechanistic feature is the formation of two covalent tetrahedral intermediates along the reaction coordinate. Given this mechanism, a

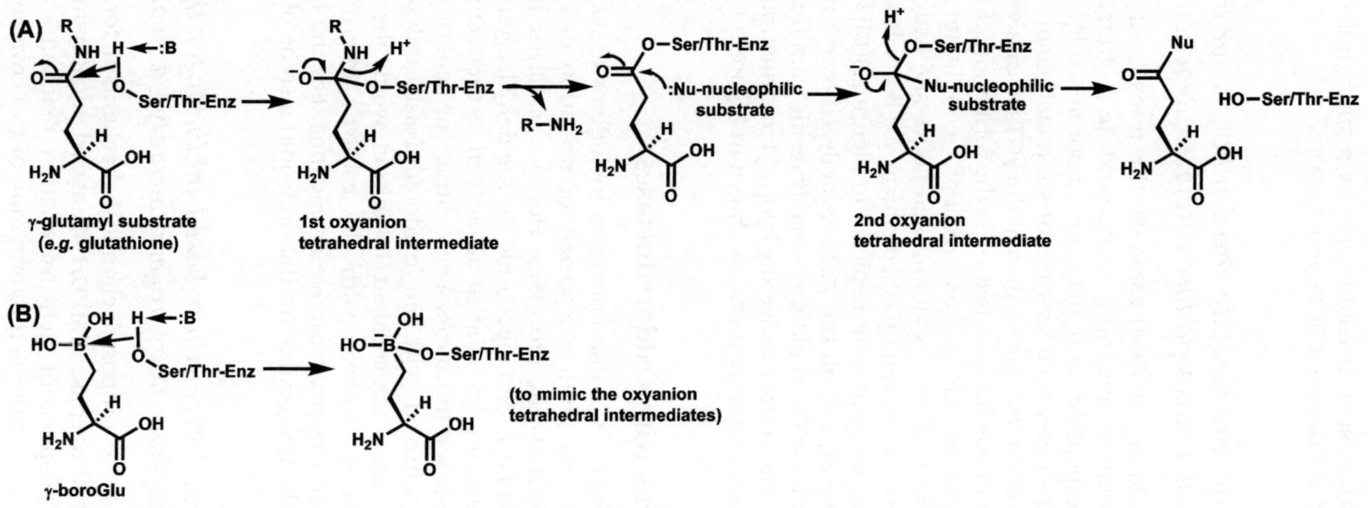

Figure 2.3 (A) The proposed chemical mechanism of the γGTase-catalyzed *trans*-γ-glutamylaton reaction. Enz, enzyme (*i.e.* γGTase); Ser, serine; Thr, threonine; Nu:, a nucleophile from the nucleophilic substrate (*e.g.* a dipeptide) bound at the active site of γGTase; B:, a general base at the active site of γGTase. (B) The chemical structure of the γ-boronic acid analog γ-boroGlu and the γGTase-catalyzed reaction of γ-boroGlu with the active site hydroxyl leading to the formation of a structural mimic of the two oxyanion intermediates depicted in (A).

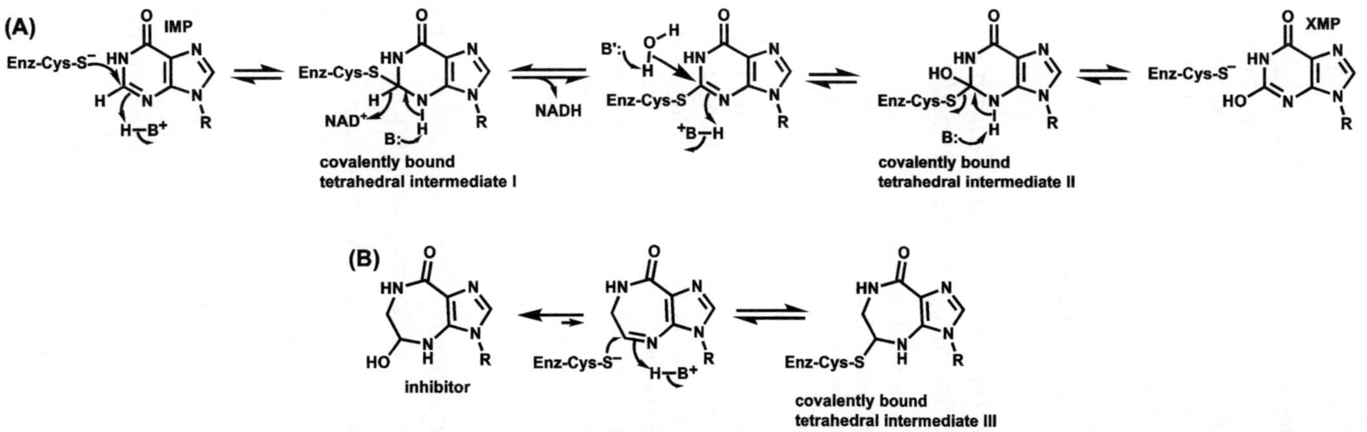

Figure 2.4 (A) The proposed chemical mechanism of the IMPDH-catalyzed NAD^+-dependent dehydrogenation reaction. Please see other figures (*e.g.* Figure 3.7) in this book for the full chemical structures of NAD^+ and NADH. B: and B′:, general bases at the active site of IMPDH. (B) The proposed IMPDH-catalyzed processing of the compound named "inhibitor" leading to the formation of the covalent intermediate III which would structurally mimic the covalent intermediates I and II depicted in (A).

so-called "fat base" nucleotide (*i.e.* the compound named "inhibitor" in Figure 2.4(B)) was designed and was found to be a potent inhibitor of the IMPDH-catalyzed reaction.[6] Figure 2.4(B) depicts the IMPDH-catalyzed processing of this inhibitory compound. Its potent inhibitory action would presumably result from the ability of its reaction product with the IMPDH active site cysteine side chain SH (*i.e.* the depicted covalent tetrahedral intermediate III) to structurally mimic the covalent tetrahedral intermediates I and II shown in Figure 2.4(A). Given the therapeutic significance of the IMPDH-catalyzed reaction, the identification of "inhibitor" as a potent inhibitor for the IMPDH-catalyzed reaction would facilitate the development of therapeutic agents for human diseases.

2.2.2.2 A Catalytic Intermediate-based Inhibitor for the Glutamate Dehydrogenase-catalyzed Reaction

Glutamate dehydrogenase (GDH) catalyzes the deamination of L-glutamate with the formation of 2-ketoglutarate, which constitutes one important metabolic reaction for amino acid catabolism. Figure 2.5(A) depicts the proposed chemical mechanism for the GDH-catalyzed reaction.[7] The key mechanistic feature is the formation of the depicted intermediate 2-iminoglutarate. Figure 2.5(B) depicts the chemical structure of 2-methyleneglutarate which is a close structural mimic of 2-iminoglutarate shown in Figure 2.5(A), and because of this, 2-methyleneglutarate could be an inhibitor for the GDH-catalyzed reaction. Indeed, this has been observed experimentally.[7] Given that all the GDHs share the common chemical mechanism depicted in Figure 2.5(A) regardless of whether NAD^+ or $NADP^+$ is used as the coenzyme for the GDH-catalyzed reaction, 2-methyleneglutarate would be a universal GDH inhibitor.

2.2.2.3 A Catalytic Intermediate-based Inhibitor for the 6-Phosphogluconate Dehydrogenase-catalyzed Reaction

6-Phosphogluconate dehydrogenase (6PGDH) catalyzes the oxidative decarboxylation of 6-phosphogluconate (6PG) with the production of NADPH and ribulose-5-phosphate (Ru5P), which is a part of the pentose phosphate shunt. Figure 2.6(A) depicts the proposed chemical mechanism for the 6PGDH-catalyzed reaction.[8] The key mechanistic feature is the formation of the high-energy 1,2-enediolate intermediate. If the CH at the C1 position of this intermediate is isosterically replaced with N, as depicted in Figure 2.6(B), the resulting compound would be a close structural analog of this intermediate, and as a result it and its stable hydroxamate tautomer would be a potent inhibitor for the 6PGDH-catalyzed reaction. Given the importance of the pentose phosphate shunt in fundamental life processes and its presence in all the evolutionary life forms, the inhibitors for the 6PGDH-catalyzed reaction could be potentially developed into therapeutic agents for human diseases if

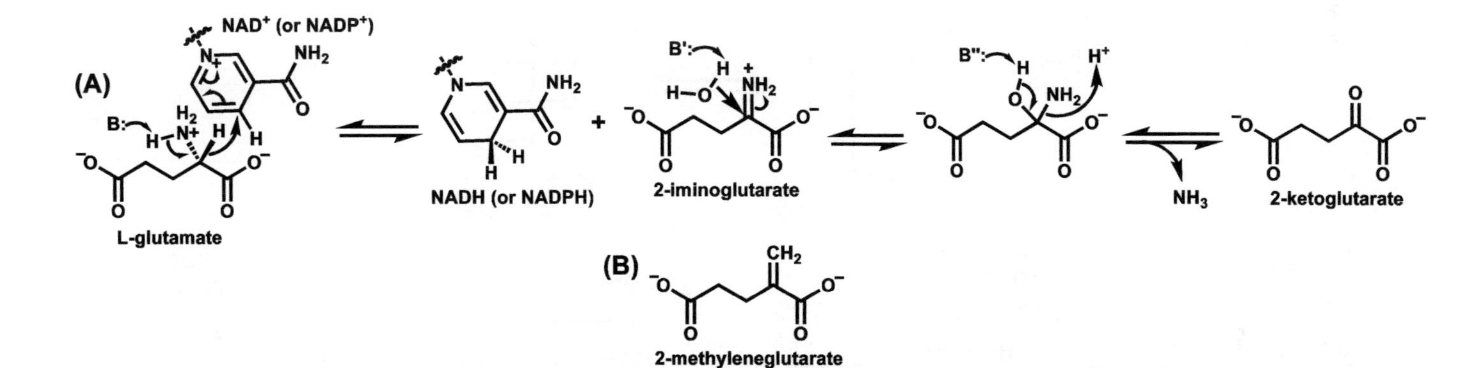

Figure 2.5 (A) The proposed chemical mechanism for the GDH-catalyzed reaction. B:, B′:, and B″:, general bases at the active site of GDH. Please see other figures (*e.g.* Figure 3.7) in this book for the full chemical structures of $NAD(P)^+$ and NAD(P)H. (B) The chemical structure of 2-methyleneglutarate which would structurally mimic 2-iminoglutarate shown in (A).

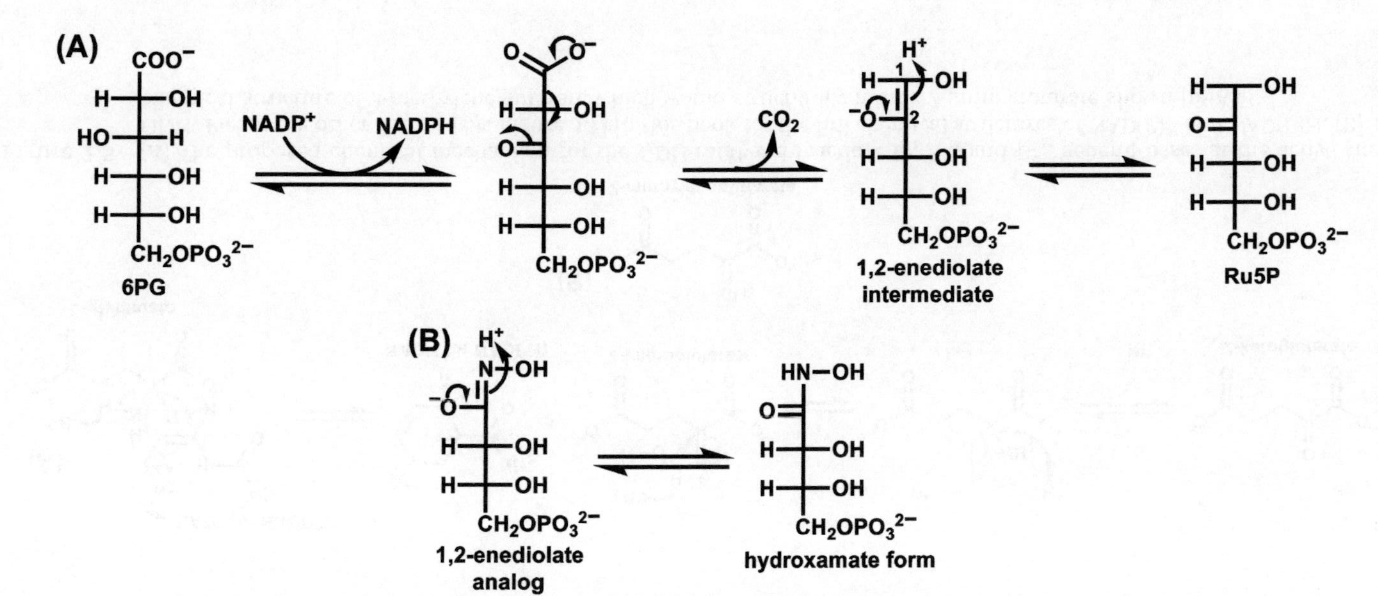

Figure 2.6 (A) The proposed chemical mechanism for the 6PGDH-catalyzed reaction. Please see other figures (*e.g.* Figure 3.7) in this book for the full chemical structures of $NADP^+$ and NADPH. (B) The chemical structures of the structural analog of the 1,2-enediolate intermediate and its stable hydroxamate tautomer.

selective 6PGDH inhibition could be realized between humans and pathogenic organisms. Indeed, it seems to be feasible to realize this potential given the successful development of the inhibitors selectively against the 6PGDH in the pathogenic parasite *Trypanosoma brucei versus* that in sheep.

2.2.3 Applications with Hydrolases

As mentioned in Chapter 1, hydrolases catalyze the hydrolytic reactions on both protein and non-protein substrates, which are involved in a variety of important life processes. The substrates can be molecules harboring scissile amide (including β-lactam), acetal, ester (including phosphate, sulfate, and thioester), or thioether linkages. The catalytic intermediate-based inhibitors have been developed for the hydrolase-catalyzed reactions on the amide and acetal substrates, as illustrated below with notable examples. Unlike the scenarios with the catalytic mechanism-based inhibitors, only non-covalent inhibition has been observed with the catalytic intermediate-based inhibitors identified for the hydrolase-catalyzed reactions.

2.2.3.1 A Catalytic Intermediate-based Inhibitor for the Carboxypeptidase A-catalyzed Amide Hydrolysis Reaction

Carboxypeptidase A (CPA) is a Zn^{2+}-dependent metalloprotease with important biological regulatory functions. It catalyzes the amino acid residue removal from the carboxyl terminus (*i.e.* C-term) of a protein substrate whose preferred C-term residues would be bulky hydrophobic residues such as phenylalanine (Phe).

Figure 2.7(A) depicts the proposed chemical mechanism for the CPA-catalyzed Zn^{2+}-dependent amide hydrolysis reaction.[9] The key mechanistic feature is the Zn^{2+}-activation of the catalytic water molecule for its subsequent nucleophilic attack on the carbonyl carbon of the scissile peptide bond, leading to the formation of the depicted tetrahedral intermediate. Figure 2.7(B) depicts the chemical structures of the aldehyde-based CPA inhibitor DL-2-benzyl-3-formylpropanoic acid (BFP) and its hydrated tetrahedral form whose structure is presumably able to mimic the catalytic tetrahedral intermediate depicted in Figure 2.7(A).[9]

2.2.3.2 A Catalytic Intermediate-based Inhibitor for the Zika Virus NS2B/NS3 Protease-catalyzed Amide Hydrolysis Reaction

Zika virus NS2B/NS3 protease is a serine protease that plays an important regulatory role in viral replication and maturation, and as such its inhibitors would exhibit an anti-viral effect. Being a serine protease, NS2B/NS3 protease would follow a covalent chemical mechanism in which the catalytic serine residue would initially nucleophilically attack the carbonyl carbon of the scissile peptide bond in a substrate molecule with the formation of an

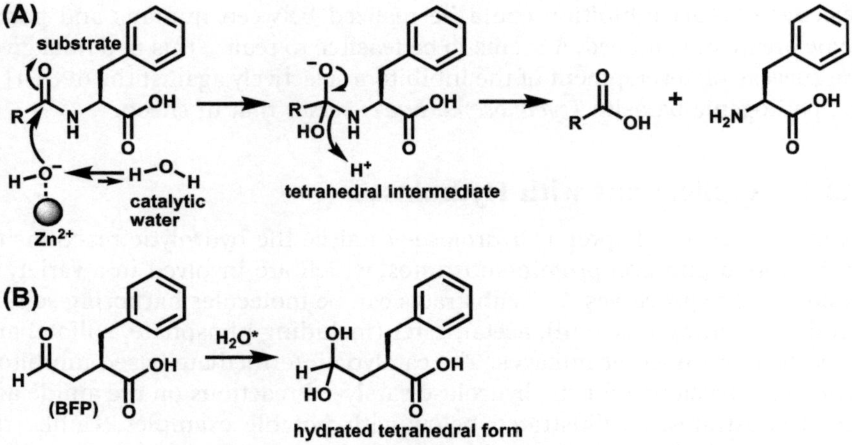

Figure 2.7 (A) The proposed chemical mechanism for the CPA-catalyzed Zn^{2+}-dependent amide hydrolysis reaction on a substrate molecule with a C-term phenylalanine residue. R, the remaining part of the substrate molecule. Zn^{2+} serves to stabilize and enrich the hydroxide (OH^-) from water dissociation (depicted) and to further polarize the carbonyl of the scissile amide bond (not depicted). (B) The chemical structures of the aldehyde-based CPA inhibitor BFP and its hydrated tetrahedral form. The latter structure is presumably able to mimic the catalytic tetrahedral intermediate depicted in (A). The asterisk in "H_2O*" indicates that this water molecule could be that in the bulk free solution or the catalytic water molecule in the CPA active site.

acylated enzyme intermediate and the C-term peptidic amine. The second half of the catalysis would be the hydrolysis of this intermediate leading to the formation of the second enzymatic reaction product, *i.e.* the N-term peptidic acid. Figure 2.8(A) depicts the proposed chemical mechanism for the NS2B/NS3 protease-catalyzed amide hydrolysis reaction.[10] Given this mechanism, the dipeptidyl aldehyde depicted in Figure 2.8(B) was designed and found to be a potent inhibitor of the NS2B/NS3 protease-catalyzed reaction. As depicted in Figure 2.8(B), the inhibitor would be converted to the depicted covalent tetrahedral intermediate following the interaction of its aldehydic group with the side chain hydroxyl of the catalytic serine residue. This tetrahedral intermediate would structurally mimic the two tetrahedral intermediates formed during the normal enzymatic catalysis as depicted in Figure 2.8(A).[10]

2.2.3.3 A Catalytic Intermediate-based Inhibitor for the α-Thrombin-catalyzed Amide Hydrolysis Reaction

As described above, peptide aldehydes constitute a fairly effective type of inhibitory warheads for the protease-catalyzed reactions owing to the ability of their hydrated (or hydroxylated) form to structurally mimic a tetrahedral

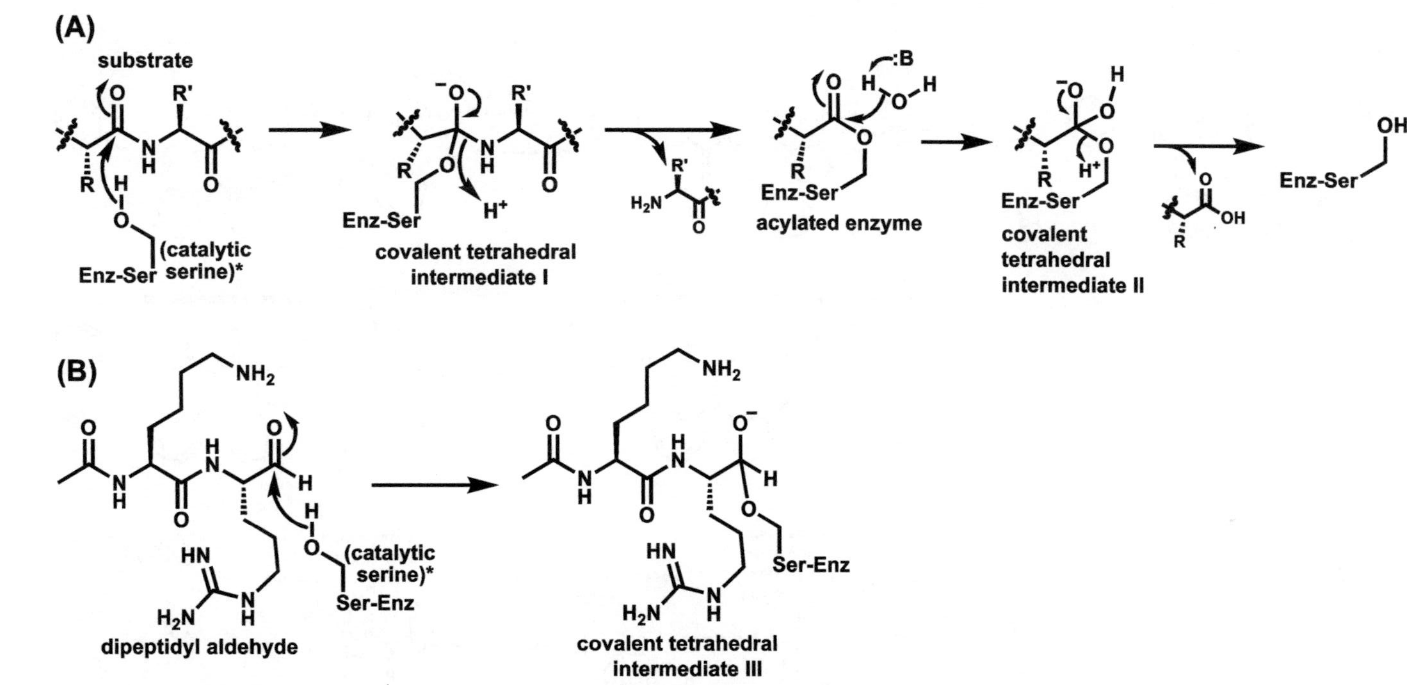

Figure 2.8 (A) The proposed chemical mechanism for the NS2B/NS3 protease-catalyzed amide hydrolysis reaction. R and R′, the side chains flanking the scissile peptide bond. (B) The NS2B/NS3 protease-catalyzed processing of the dipeptidyl aldehyde inhibitor leading to the formation of the covalent tetrahedral intermediate III which would structurally mimic the covalent tetrahedral intermediates I and II depicted in (A). Enz, enzyme (*i.e.* the NS2B/NS3 protease); B:, a general base at the active site of the NS2B/NS3 protease. The asterisk in "(catalytic serine)*" indicates that its side chain hydroxyl would be activated by histidine and aspartate residues (the other two residues of the catalytic triad; not depicted) before the nucleophilic attack.

(A)
substrate
(catalytic serine)*
Enz-Ser
covalent tetrahedral intermediate I
acylated enzyme
covalent tetrahedral intermediate II

(B)
phosphonate tripeptide
(catalytic water)*
pentacoordinated phosphorus intermediate
(catalytic serine)*
Enz-Ser
tetracoordinated phosphorus intermediate

intermediate formed during normal enzymatic catalysis. In addition to the aldehydic functionality, phosphonate has also been found to be an effective inhibitory warhead for the protease-catalyzed reactions, for example for the serine protease α-thrombin-catalyzed reaction as detailed below, owing to the ability of its hydroxylated tetracoordinated phosphorus intermediate to structurally mimic a tetrahedral intermediate formed during normal enzymatic catalysis.

Figure 2.9(A) depicts the proposed chemical mechanism for the α-thrombin-catalyzed amide hydrolysis reaction, which is the same as that for the NS2B/NS3 protease-catalyzed amide hydrolysis reaction depicted in Figure 2.8(A), since they are both serine proteases. Figure 2.9(B) depicts the α-thrombin-catalyzed processing of a phosphonate tripeptidic inhibitor for the α-thrombin-catalyzed amide hydrolysis reaction, leading to the formation of the depicted tetracoordinated phosphorus intermediate (*via* the depicted pentacoordinated phosphorus intermediate) which would structurally mimic the covalent tetrahedral intermediates I and II depicted in Figure 2.9(A).[11] It is worth noting that the formation of the depicted pentacoordinated phosphorus intermediate and tetracoordinated phosphorus intermediate has been corroborated by X-ray crystallographic analysis in which they were directly observed from the incubations of the phosphonate tripeptidic inhibitor with α-thrombin for a shorter and a longer period of time, respectively.

2.2.3.4 A Catalytic Intermediate-based Inhibitor for the Glycosidase HypBA1-catalyzed Acetal Hydrolysis Reaction

The glycosidase-catalyzed reactions play an important regulatory role in multiple important life processes by virtue of their ability to counteract the glycosyltransferase-catalyzed glycosylation of a variety of biomolecules including biomacromolecules. It was recently found that certain glycosidases have a catalytic dependence on Zn^{2+}, and in order to facilitate a mechanistic elucidation and explore the therapeutic potential of the inhibition of the reactions catalyzed by these newly discovered glycosidases, potent inhibitors are needed. Figure 2.10(A) and (B) depict the proposed chemical mechanism for

Figure 2.9 (A) The proposed chemical mechanism for the α-thrombin-catalyzed amide hydrolysis reaction. R and R′, the side chains flanking the scissile peptide bond. This scheme is identical to that depicted in Figure 2.8(A). (B) The α-thrombin-catalyzed processing of the phosphonate tripeptidic inhibitor leading to the formation of the depicted pentacoordinated phosphorus intermediate and tetracoordinated phosphorus intermediate. The latter would structurally mimic the covalent tetrahedral intermediates I and II depicted in (A). Enz, enzyme (*i.e.* α-thrombin). The asterisks in "(catalytic water)*" and "(catalytic serine)*" indicate that this water molecule and the side chain hydroxyl of the catalytic serine residue could be both activated by un-depicted histidine and aspartate residues (the other two residues of the catalytic triad) before the nucleophilic attack.

this unique family of glycosidase enzymes and the catalytic processing of an inhibitor depicted as "β-L-arabinofuranosyl epoxide", respectively.[12] Due to the greater stability of the C–S bond than the C–O bond encountered with other glycosidases, the biological logic of this family of enzymes for a catalytic dependence on Zn^{2+} could be appreciated as one way to facilitate the C–S bond cleavage during the dethioglycosylation step of the enzymatic reaction, as depicted in Figure 2.10(A). In addition, Zn^{2+} would also activate the side chain thiol (SH) of the catalytic cysteine residue by stabilizing and enriching the thiolate anion (S^-) for the nucleophilic attack on the anomeric carbon of the glycoside substrate. As depicted in Figure 2.10(B), the HypBA1-catalyzed processing of β-L-arabinofuranosyl epoxide would also lead to the formation of a thioglycosyl enzyme intermediate (tGEI), however, this tGEI would be unable to be dethioglycosylated, because its ring structure is the carbocyclic cyclopentane ring instead of the furanosyl ring present in the substrate molecule.

2.2.4 Applications with Lyases

As discussed with notable examples in Chapter 1, catalytic mechanism-based inhibitors have been developed for the lyation (group elimination) reactions catalyzed by lyases and other enzymes (*e.g.* decarboxylase) which play important regulatory roles in a variety of important life processes. In this section, notable examples of the catalytic intermediate-based inhibitors for the enzymatic lyation reactions are discussed. Unlike the scenarios with the above-described catalytic mechanism-based inhibitors, only non-covalent inhibition has been observed with the catalytic intermediate-based inhibitors identified for the enzymatic lyation reactions.

2.2.4.1 A Catalytic Intermediate-based Inhibitor for the Tryptophan Indole-lyase-catalyzed Reaction

Tryptophan indole-lyase (TIL) catalyzes the pyridoxal-5′-phosphate (PLP)-dependent C_β–C_γ cleavage on tryptophan (L-Trp) side chain with the formation of free indole, pyruvate, and ammonium cation. This enzymatic reaction occurs only in bacteria and plays an important role in regulating bacterial pathogenesis and antibiotic resistance *via* the signaling role of free indole. Thus, the inhibitors of the TIL-catalyzed reaction are potential antibiotics.

Figure 2.11(A) depicts the proposed chemical mechanism for the TIL-catalyzed PLP-dependent lyation reaction.[13] The key mechanistic step is the depicted C_β–C_γ cleavage on the quinonoid intermediate generating the indole ring with a negative charge on the C-3 carbon, whose driving force is the stabilization of such a negative charge *via* electron delocalization on to the aromatic bicyclic ring system. The TIL-catalyzed PLP-dependent processing of L-bishomotryptophan (a L-Trp analog) is shown in Figure 2.11(B), however, unlike normal enzymatic catalysis, the formed external aldimine intermediate II was found not to be further converted to the corresponding

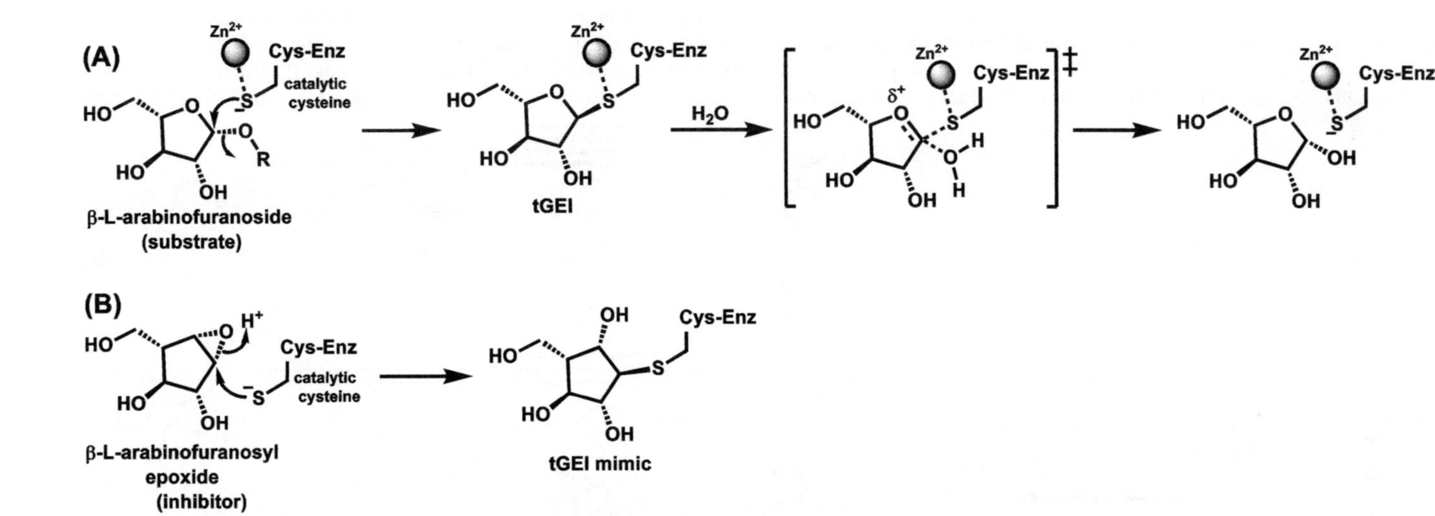

Figure 2.10 (A) The proposed chemical mechanism for the glycoside hydrolysis of a β-L-arabinofuranoside substrate catalyzed by the Zn^{2+}-dependent glycosidase HypBA1. R, the remainder of the substrate molecule; tGEI, thioglycosyl enzyme intermediate. (B) The HypBA1-catalyzed processing of β-L-arabinofuranosyl epoxide (an inhibitor) leading to the formation of the depicted thioglycosylated enzyme intermediate (a tGEI mimic). Enz, enzyme (*i.e.* HypBA1); Cys, cysteine.

(A)

PLP

TIL

Covalently anchored PLP to TIL (*via* the depicted protonated Schiff base)

L-Trp

Covalently linked PLP to L-Trp (*via* the depicted protonated Schiff base) external aldimine intermediate I

quinonoid intermediate

stabilized *via* electron delocalization

H_2O

Indole

PLP

NH_4^+

(B)

PLP

TIL

Covalently anchored PLP to TIL (*via* the depicted protonated Schiff base)

L-bishomo-tryptophan

Covalently linked PLP to L-bishomotryptophan (*via* the depicted protonated Schiff base) external aldimine intermediate II

Figure 2.11 (A) The proposed chemical mechanism for the TIL-catalyzed PLP-dependent lyation reaction on L-Trp. TIL-B:, a general base at the TIL active site. (B) The TIL-catalyzed PLP-dependent processing of L-bishomotryptophan leading to the formation of external aldimine intermediate II which would structurally mimic external aldimine intermediate I depicted in (A) that is formed during the normal TIL catalysis on L-Trp.

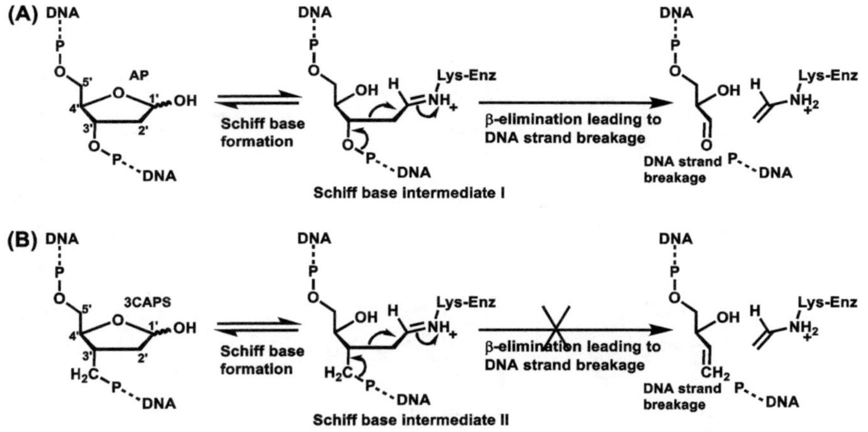

Figure 2.12 (A) The proposed chemical mechanism for the AP lyase activity of the bi-functional glycosylases on an AP-site DNA substrate. (B) The lyase-catalyzed processing of a 3CAPS-containing DNA molecule leading to the formation of Schiff base intermediate II which would structurally mimic Schiff base intermediate I depicted in (A) that is formed during the normal lyase catalysis on an AP-site DNA substrate. Enz, enzyme (*i.e.* the bi-functional glycosylase); Lys, active site lysine residue.

quinonoid intermediate.[13] Nevertheless, external aldimine intermediate II would structurally mimic external aldimine intermediate I formed during the normal TIL catalysis on L-Trp and shown in Figure 2.11(A), rendering L-bishomotryptophan a potent catalytic intermediate-based inhibitor for the TIL-catalyzed lyation reaction. Of note, the inability of external aldimine intermediate II to be converted to the corresponding quinonoid intermediate could be due to the lack of an appropriately positioned general base (*i.e.* TIL-B:) to perform the depicted α-deprotonation following the binding of L-bishomotryptophan at the TIL active site.

2.2.4.2 A Catalytic Intermediate-based Inhibitor for the Bifunctional DNA Glycosylase-catalyzed AP Lyation Reaction

Base excision repair constitutes an important mechanism for DNA damage repair and involves the excision of a mismatched or damaged nucleobase from a DNA duplex, leaving behind an abasic site (AP-site). Various glycosylase enzymes play an essential role during this process, and one particular subset of these glycosylases are bi-functional in that they also possess an AP lyase activity, *i.e.* catalyzing DNA strand breakage at an AP-site.

Figure 2.12(A) depicts the proposed chemical mechanism for this AP lyase activity.[14] The key mechanistic feature is the formation of the depicted sugar ring-opened protonated Schiff base (between a lysine side chain from

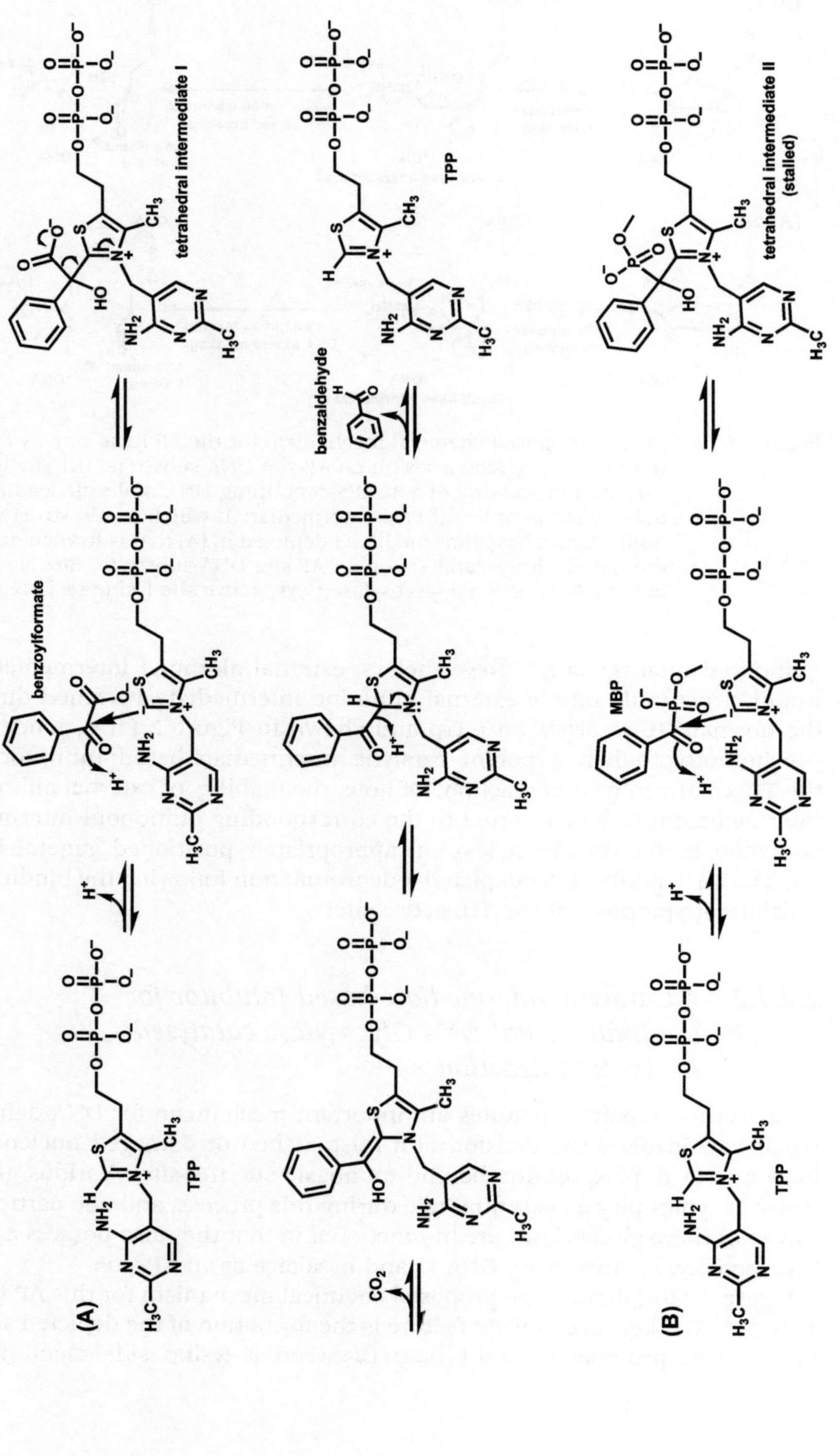

(A)

TPP

benzoylformate

H^+

CO_2

tetrahedral intermediate I

benzaldehyde

TPP

(B)

TPP

MBP

H^+

tetrahedral intermediate II (stalled)

enzyme active site and the anomeric carbon (C1′) at the AP-site) which would serve as an electron sink promoting the subsequent DNA strand breakage.

By replacing the AP-site 3′-oxygen (O) with methylene (CH_2) in an AP-site DNA substrate, the resulting DNA molecule harboring the analog (*i.e.* 3′-carbon AP-site or 3CAPS) was found to be converted by the lyase activity to the corresponding protonated Schiff base, as depicted in Figure 2.12(B), however, due to the presence of a P–C bond (of a phosphonate linkage) instead of a P–O bond (of a phosphate linkage) at the 3′-position, the further transformation of Schiff base intermediate II would be impossible. However, this intermediate would structurally mimic Schiff base intermediate I formed during normal enzymatic catalysis as shown in Figure 2.12(A), giving rise to an inhibition of the AP lyase activity.[15]

2.2.4.3 A Catalytic Intermediate-based Inhibitor for the Benzoylformate Decarboxylase-catalyzed Reaction

Benzoylformate decarboxylase (BFDC) catalyzes the thiamine pyrophosphate (TPP)-dependent decarboxylation of benzoylformate to form benzaldehyde, which plays an important regulatory role in metabolism. Figure 2.13(A) depicts the proposed chemical mechanism for this decarboxylation reaction.[16] The key mechanistic step is the formation of the depicted tetrahedral intermediate I before the decarboxylation step. Figure 2.13(B) depicts the BFDC-catalyzed TPP-dependent processing of methyl benzoylphosphonate (MBP), an analog of the native substrate benzoylformate.[16] As depicted, BFDC is also able to catalyze the formation of a covalent conjugate of MBP with coenzyme TPP (*i.e.* tetrahedral intermediate II), however, this intermediate is not able to be chemically transformed further, and thus it gets stalled once formed at the BFDC active site, and because of this, the enzymatic activity of BFDC consequently gets inhibited. Since tetrahedral intermediate II would structurally mimic tetrahedral intermediate I formed during normal enzymatic catalysis as depicted in Figure 2.13(A), MBP could be regarded as a catalytic intermediate-based inhibitor.

2.2.5 Applications with Isomerases

As stated in Chapter 1, isomerases catalyze a variety of reactions involved in various important life processes. In this section, the development of the catalytic intermediate-based inhibitors for the isomerase-catalyzed reaction and for the reaction catalyzed by a racemase is illustrated with notable

Figure 2.13 (A) The proposed chemical mechanism for the BFDC-catalyzed TPP-dependent decarboxylation of benzoylformate to form benzaldehyde. (B) The BFDC-catalyzed TPP-dependent processing of MBP with the formation of the stalled tetrahedral intermediate II which would structurally mimic tetrahedral intermediate I formed during normal enzymatic catalysis as depicted in (A).

examples. Unlike the scenario with the above-described catalytic mechanism-based inhibitors, only non-covalent inhibition has been observed with the catalytic intermediate-based inhibitors identified for the enzymatic isomerization and racemization reactions.

2.2.5.1 A Catalytic Intermediate-based Inhibitor for the Phosphoglucose Isomerase-catalyzed Reaction

Phosphoglucose isomerase (PGI) catalyzes the inter-conversion between D-glucose-6-phosphate (G6P) and D-fructose-6-phosphate (F6P), as depicted in Figure 2.14(A). PGI is a glycolytic enzyme and as such the PGI-catalyzed reaction plays an important role in regulating central energy metabolism. Additionally, the PGI-catalyzed reaction is pathologically significant particularly concerning its involvement in the central energy metabolism in tumors and various parasitic pathogens. Therefore, the inhibitors for the PGI-catalyzed reaction would constitute an important class of therapeutically important compounds.

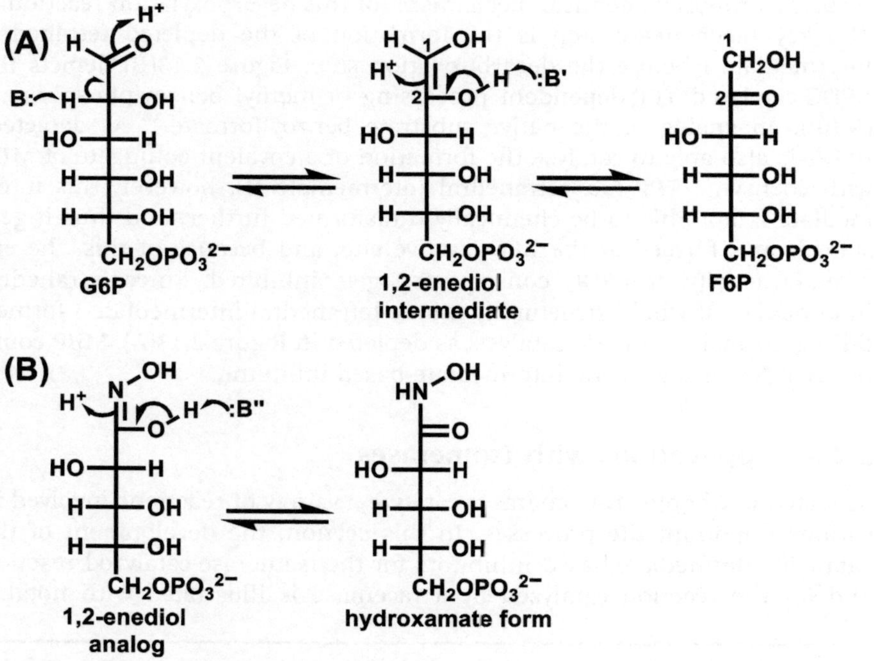

Figure 2.14 (A) The proposed chemical mechanism for the PGI-catalyzed reaction. B: and B′:, general bases at the PGI active site. (B) The chemical structures of a close structural analog of the 1,2-enediol intermediate shown in (A) and its stable hydroxamate tautomer. B″:, a general base at the PGI active site.

Figure 2.14(A) depicts the proposed chemical mechanism for the PGI-catalyzed reaction.[17] The key mechanistic feature is the formation of the depicted high-energy 1,2-enediol intermediate. Figure 2.14(B) shows the chemical structures of a close structural analog of this catalytic intermediate (*i.e.* that derived from the isosteric replacement of N for CH at the C1 position of the 1,2-enediol intermediate) and its stable hydroxamate tautomer. Being able to structurally mimic the catalytic enediol intermediate, this analog could be a potent inhibitor for the PGI-catalyzed reaction, as indeed has been observed experimentally.[17]

2.2.5.2 A Catalytic Intermediate-based Inhibitor for the Phosphomannose Isomerase-catalyzed Reaction

Phosphomannose isomerase (PMI) catalyzes the inter-conversion between D-mannose 6-phosphate (M6P) and D-fructose 6-phosphate (F6P), the same type of reaction as the above-described PGI-catalyzed reaction. However, while PGI is not a metalloenzyme, all the currently known PMIs have an absolute catalytic dependence on divalent metal cations such as Zn^{2+}.

Figure 2.15(A) depicts the proposed chemical mechanism for the PMI-catalyzed metal cation-dependent reaction.[18] Same as the PGI-catalyzed reaction, the key mechanistic feature of the PMI-catalyzed reaction is the formation of a high-energy 1,2-enediol catalytic intermediate which is otherwise stabilized *via* chelation with a divalent metal cation in the case with the PMI-catalyzed reaction, as depicted in Figure 2.15(A).

Figure 2.15(B) depicts the chemical structures of a close structural analog of the 1,2-enediol intermediate (*i.e.* the one derived from the isosteric replacement of N for CH at the C1 position of the 1,2-enediol intermediate) and its stable hydroxamate tautomer. Of note, this analog is same as that described in Section 2.2.5.1. Being able to structurally mimic the catalytic enediol intermediate and to additionally form a chelate with the metal cation at enzyme active site, this analog has been observed experimentally to be a potent inhibitor for the PMI-catalyzed reaction.[18] Given the metabolic and therapeutic importance of the PMI-catalyzed reaction, particularly regarding human pathogens and their chemotherapy, this analog could be a valuable lead for further development.

2.2.5.3 A Catalytic Intermediate-based Inhibitor for the Mandelate Racemase-catalyzed Reaction

Mandelate racemase (MR) catalyzes the inter-conversion of (*S*)- and (*R*)-mandelate, which is a reaction useful as a model for a mechanistic understanding of the biochemical proton transfer. Figure 2.16(A) depicts the proposed chemical mechanism for the MR-catalyzed racemization reaction.[19] The key mechanistic feature is the deprotonation/reprotonation at

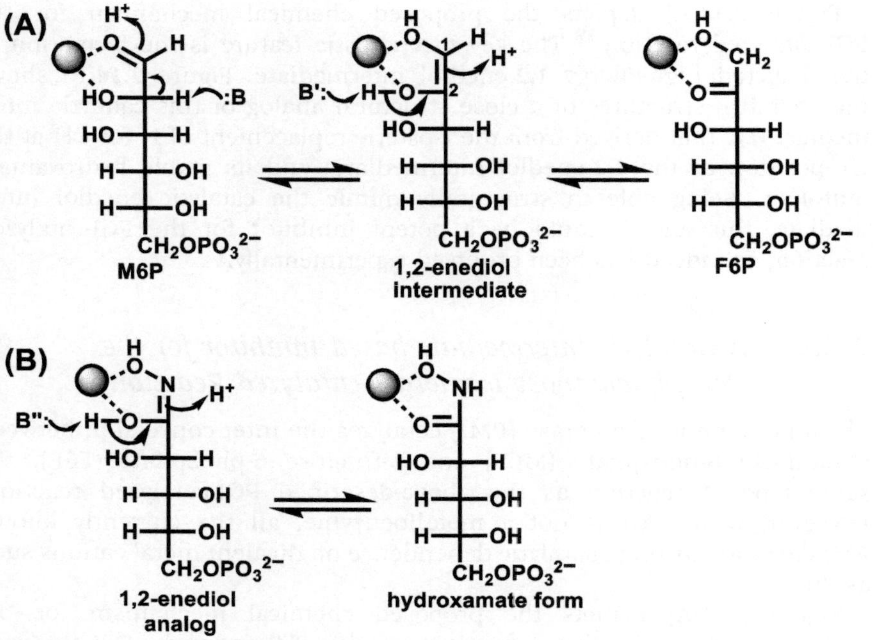

Figure 2.15 (A) The proposed chemical mechanism for the PMI-catalyzed reaction. B: and B′:, general bases at PMI active site; ● denotes a divalent metal cation (*e.g.* Zn^{2+}). (B) The chemical structures of a close structural analog of the 1,2-enediol intermediate shown in (A) and its stable hydroxamate tautomer.

the acidic carbon center in the substrate, with the depicted enetriolate species as the catalytic intermediate. Figure 2.16(B) and (C) depict the chemical structures of *N*-hydroxyformanilide (HFA) and an *N*-nitroso hydroxylamine known as cupferron, respectively, whose zwitterionic forms would both be close structural analogs of the enetriolate intermediate shown in Figure 2.16(A). As a result, HFA and cupferron would both behave as potent inhibitors for the MR-catalyzed racemization reaction, which has been observed experimentally.[19]

2.2.6 Applications with Ligases

As stated in Chapter 1, ligases catalyze a variety of ligation reactions involved in important life processes. While catalytic mechanism-based inhibitors have not yet been developed for ligase-catalyzed reactions, several catalytic intermediate-based inhibitors have been developed, as detailed herein. Like the scenario with the above-described catalytic intermediate-based inhibitors, only non-covalent inhibition has been observed with the catalytic intermediate-based inhibitors identified for the enzymatic ligation reactions.

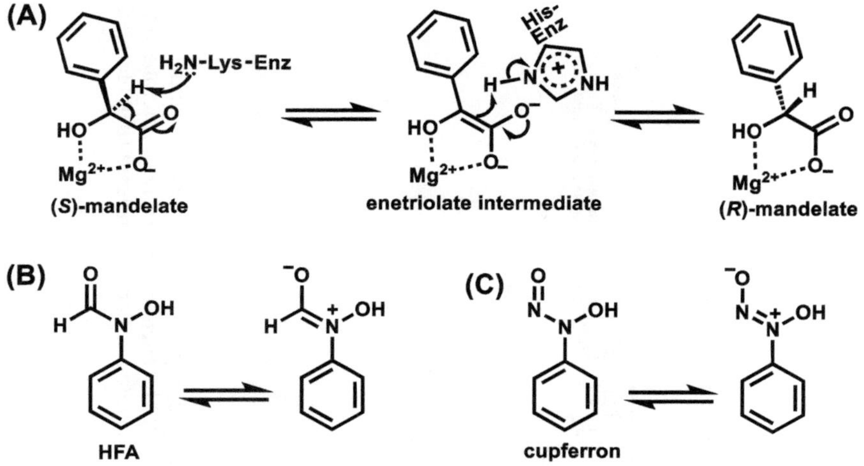

Figure 2.16 (A) The proposed chemical mechanism for the MR-catalyzed reaction. Enz, enzyme (*i.e.* MR); Lys, active site lysine residue; His, active site histidine residue. (B) The chemical structure of HFA. (C) The chemical structure of cupferron. Note: the zwitterionic forms of HFA and cupferron would both be close structural analogs of the enetriolate intermediate shown in (A).

2.2.6.1 A Catalytic Intermediate-based Inhibitor for the Dihydroxybenzoate-AMP Ligase-catalyzed Reaction

Dihydroxybenzoate-AMP ligase (EntE) catalyzes the ATP-dependent condensation between 2,3-dihydroxybenzoic acid (DHB) and phosphopantetheinylated EntB with the formation of the phosphopantetheine-crosslinked DHB and EntB (Figure 2.17(A)), which is a vital product for the final assembly of the siderophore enterobactin (Figure 2.17(B)) in *Escherichia coli*.[20] Of note, EntE is one of six proteins encoded by the enterobactin synthetase gene cluster. Given the importance of the ability of a pathogenic bacterium such as *E. coli* to synthesize a siderophore like enterobactin to its (patho)physiological functions and the vital role of the EntE-catalyzed reaction for the biosynthesis of enterobactin, the inhibitors of the EntE-catalyzed reaction could be developed into antibacterial agents.

Figure 2.17(A) depicts the proposed chemical mechanism for the EntE-catalyzed reaction.[20] The key mechanistic step is the activation of the free carboxyl group of DHB in the form of a high-energy phosphoanhydride linkage as in the depicted intermediate "adenylated DHB", which is able to be converted to another high-energy linkage, *i.e.* the thioester linkage in the reaction product. It should be noted that the driving force for the formation of the phosphoanhydride-containing high-energy intermediate is the pyrophosphatase (PPase)-catalyzed hydrolysis of pyrophosphate (PPi) to inorganic phosphate (Pi), as depicted in Figure 2.17(A).

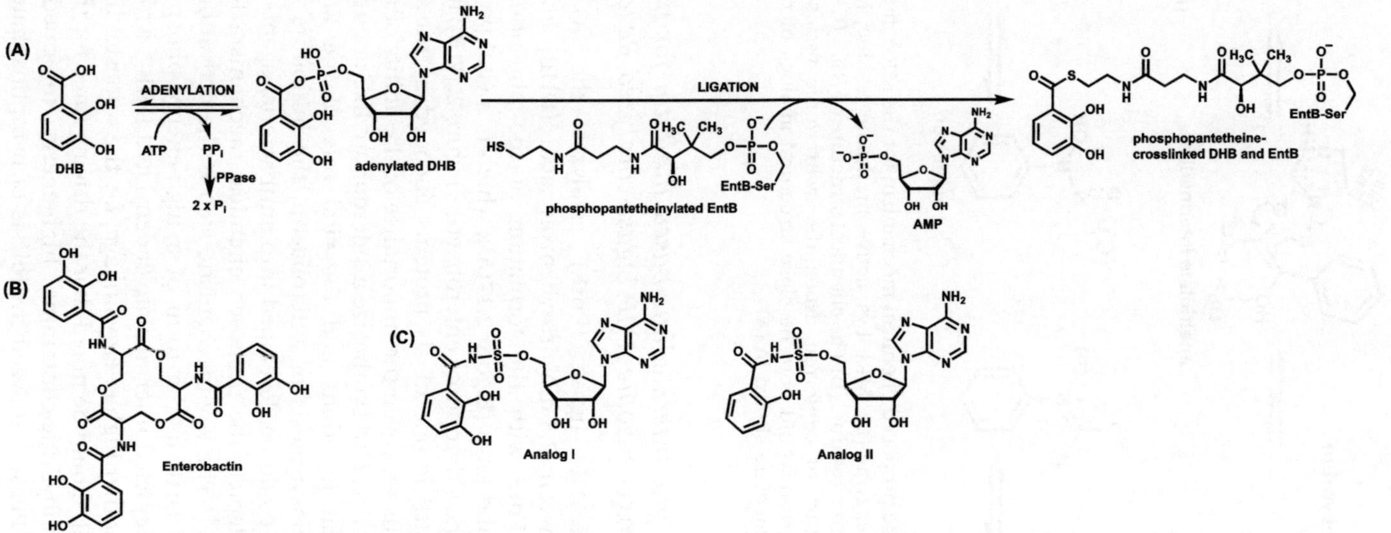

Figure 2.17 (A) The proposed chemical mechanism for the EntE-catalyzed ATP-dependent ligation reaction. Ser, serine. Note: the first reaction in this scheme is driven by the PPase-catalyzed hydrolysis of PPi to Pi. (B) The chemical structure of enterobactin. (C) The chemical structures of the two analogs which would structurally mimic the intermediate "adenylated DHB" formed during normal enzymatic catalysis as depicted in (A).

Figure 2.17(C) depicts the chemical structures of the two close structural analogs of "adenylated DHB" depicted in Figure 2.17(A). Due to the close structural mimicry with the catalytic intermediate (*i.e.* "adenylated DHB") formed during normal enzymatic catalysis, these two analogs would also be able to bind to EntE active site, however, they were not able to be further chemically transformed, therefore, they both behave as potent inhibitors for the EntE-catalyzed reaction.[20]

2.2.6.2 A Catalytic Intermediate-based Inhibitor for the tRNA-dependent Ligase MurM-catalyzed Reaction

The ligase MurM from the highly penicillin resistant strains of *Streptococcus pneumoniae* is able to catalyze the addition of an alanine residue (Ala) from alanyl-transfer RNA (alanyl-tRNA) on to the side chain ε-amino group of a lysine residue (Lys) in the lipid intermediate II, as shown in Figure 2.18(A), which is a key step for the formation of the Ala-Ala inter-strand cross-link found in the peptidoglycan layer of highly penicillin-resistant strains of *S. pneumoniae*.[21] Therefore, the inhibitors for the MurM-catalyzed reaction could be used synergistically with penicillin against the highly penicillin-resistant strains of *S. pneumoniae*.

Figure 2.18(A) depicts the proposed chemical mechanism for the MurM-catalyzed reaction.[21] The key mechanistic feature is the formation of the depicted tetrahedral intermediate. Figure 2.18(B) depicts the chemical structure of the 2′-deoxyadenosine-based 3′-phosphonate analog which would mimic the tetrahedral intermediate formed during the normal enzymatic catalysis as depicted in Figure 2.18(A), leading to an inhibition of the MurM-catalyzed reaction. Indeed, this compound was found experimentally to behave as a reasonably potent inhibitor of the MurM-catalyzed reaction.[21]

2.2.6.3 A Catalytic Intermediate-based Inhibitor for the Glutathionylspermidine Synthetase-catalyzed Reaction

Glutathionylspermidine synthetase (GspS) catalyzes the regio-specific condensation between the redox cofactor glutathione (GSH) and the polyamine spermidine with the formation of glutathionylspermidine (Gsp), as shown in Figure 2.19(A). The GspS-catalyzed reaction constitutes the penultimate step in the biosynthetic pathway of the endogenous antioxidant trypanothione (TSH; see Figure 2.19(B) for chemical structure) in parasites. Therefore, inhibitors of the GspS-catalyzed reaction would possess an anti-parasitic potential.[22–24]

Figure 2.19(A) depicts the proposed chemical mechanism for the GspS-catalyzed regio-specific amide bond formation between GSH and spermidine.[22–24] While the regio-specificity of the reaction would be made possible by the relative positioning of the substrate molecules (*i.e.* GSH, ATP, and spermidine) at the GspS active site, one key feature of the proposed

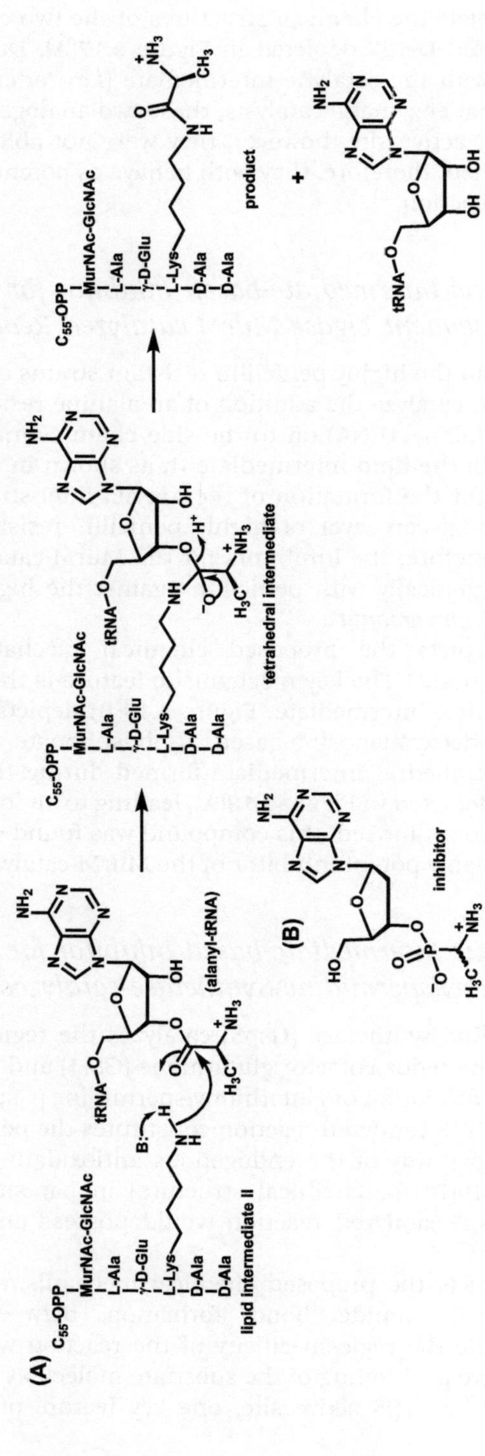

Figure 2.18 (A) The proposed chemical mechanism for the MurM-catalyzed tRNA-dependent reaction. C_{55}-OPP, undecaprenyl diphosphate; B:, a general base at MurM active site. (B) The chemical structure of the 2′-deoxyadenosine-based 3′-phosphonate inhibitor of the MurM-catalyzed reaction, which would structurally mimic the tetrahedral intermediate formed during the normal enzymatic catalysis as depicted in (A).

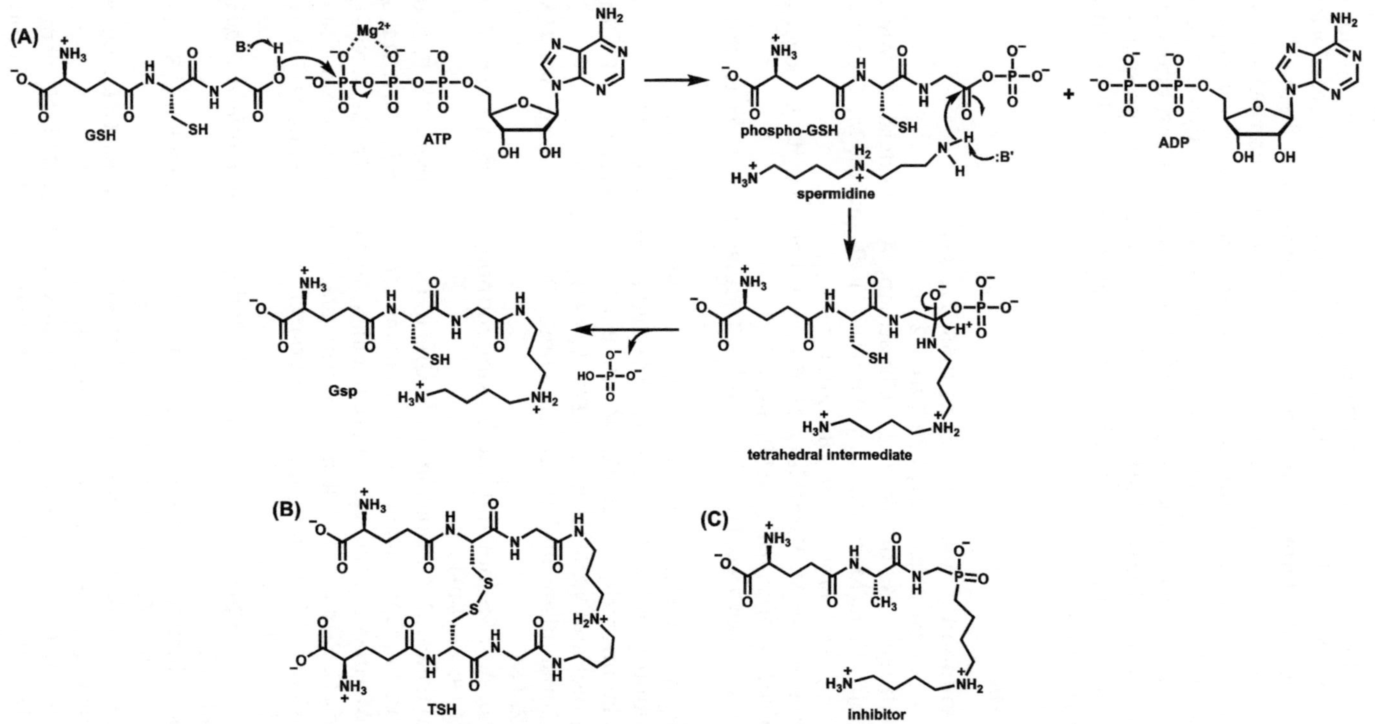

Figure 2.19 (A) The proposed chemical mechanism for the GspS-catalyzed ATP-dependent region-specific amide bond formation between GSH and spermidine. B: and B′:, general bases at the GspS active site. (B) The chemical structure of the endogenous antioxidant TSH present in parasites. (C) The chemical structure of the phosphinate inhibitor of the GspS-catalyzed reaction, which would structurally mimic the tetrahedral intermediate formed during the normal enzymatic catalysis as depicted in (A).

chemical mechanism is the activation of the depicted free carboxyl group on GSH with ATP in the form of a high-energy phosphoanhydride form; the subsequent reaction of phospho-GSH with spermidine would proceed *via* the depicted tetrahedral intermediate. Figure 2.19(C) depicts the chemical structure of a phosphinate analog of this catalytic tetrahedral intermediate, and because of its close structural mimicry with the intermediate, this compound would be expected to be a decent inhibitor of the GspS-catalyzed reaction, which has been observed experimentally.[22–24] Of note, this inhibitory compound harbors an alanine residue in place of the cysteine residue present in the tetrahedral intermediate, presumably being able to enhance the chemical stability of the compound.

References

1. G. S. Hammond, *J. Am. Chem. Soc.*, 1955, **77**, 334.
2. A. Yarnell, *Chem. Eng. News*, 2003, **81**, 42.
3. G. Sheikhnejad, A. Brank, J. K. Christman, A. Goddard, E. Alvarez, H. Ford Jr., V. E. Marquez, C. J. Marasco, J. R. Sufrin, M. O'gara and X. Cheng, *J. Mol. Biol.*, 1999, **285**, 2021.
4. S. E. Greasley, M. M. Yamashita, H. Cai, S. J. Benkovic, D. L. Boger and I. A. Wilson, *Biochemistry*, 1999, **38**, 16783.
5. R. L. Stein, C. DeCicco, D. Nelson and B. Thomas, *Biochemistry*, 2001, **40**, 5804.
6. W. Wang and L. Hedstrom, *Biochemistry*, 1998, **37**, 11949.
7. R. Choudhury and N. S. Punekar, *FEBS Lett.*, 2007, **581**, 2733.
8. R. Sundaramoorthy, J. Iulek, M. P. Barrett, O. Bidet, G. F. Ruda, I. H. Gilbert and W. N. Hunter, *FEBS J.*, 2007, **274**, 275.
9. D. W. Christianson and W. N. Lipscomb, *Proc. Natl. Acad. Sci. U.S.A.*, 1985, **82**, 6840.
10. B. Nutho, A. J. Mulholland and T. Rungrotmongkol, *Phys. Chem. Chem. Phys.*, 2019, **21**, 14945.
11. E. Skordalakes, G. G. Dodson, D. S. Green, C. A. Goodwin, M. F. Scully, H. R. Hudson, V. V. Kakkar and J. J. Deadman, *J. Mol. Biol.*, 2001, **311**, 549.
12. N. G. S. McGregor, J. Coines, V. Borlandelli, S. Amaki, M. Artola, A. Nin-Hill, D. Linzel, C. Yamada, T. Arakawa, A. Ishiwata, Y. Ito, G. A. van der Marel, J. D. C. Codée, S. Fushinobu, H. S. Overkleeft, C. Rovira and G. J. Davies, *Angew. Chem. Int. Ed.*, 2021, **60**, 5754.
13. Q. T. Do, G. T. Nguyen, V. Celis and R. S. Phillips, *Arch. Biochem. Biophys.*, 2014, **560**, 20.
14. A. K. McCullough, A. Sanchez, M. L. Dodson, P. Marapaka, J. S. Taylor and R. S. Lloyd, *Biochemistry*, 2001, **40**, 561.
15. D. Schuermann, S. P. Scheidegger, A. R. Weber, M. Bjørås, C. J. Leumann and P. Schär, *Nucleic Acids Res.*, 2016, **44**, 2187.
16. G. S. Brandt, M. M. Kneen, S. Chakraborty, A. T. Baykal, N. Nemeria, A. Yep, D. I. Ruby, G. A. Petsko, G. L. Kenyon, M. J. McLeish, F. Jordan and D. Ringe, *Biochemistry*, 2009, **48**, 3247.

17. R. Hardré, C. Bonnette, L. Salmon and A. Gaudemer, *Bioorg. Med. Chem. Lett.*, 1998, **8**, 3435.
18. C. Roux, J. H. Lee, C. J. Jeffery and L. Salmon, *Biochemistry*, 2004, **43**, 2926.
19. J. R. Bourque, R. K. M. Burley and S. L. Bearne, *Bioorg. Med. Chem. Lett.*, 2007, **17**, 105.
20. A. L. Sikora, D. J. Wilson, C. C. Aldrich and J. S. Blanchard, *Biochemistry*, 2010, **49**, 3648.
21. E. Cressina, A. J. Lloyd, G. De Pascale, B. J. Mok, S. Caddick, D. I. Roper, C. G. Dowson and T. D. H. Bugg, *Bioorg. Med. Chem.*, 2009, **17**, 3443.
22. S. L. Oza, S. Chen, S. Wyllie, J. K. Coward and A. H. Fairlamb, *FEBS J.*, 2008, **275**, 5408.
23. S. Chen, C.-H. Lin, C. T. Walsh and J. K. Coward, *Bioorg. Med. Chem. Lett.*, 1997, **7**, 505.
24. C. H. Lin, S. Chen, D. S. Kwon, J. K. Coward and C. T. Walsh, *Chem. Biol.*, 1997, **4**, 859.

CHAPTER 3

Substrate-based Design

3.1 Mode of Working

If close structural analogs of one or more substrates of an enzyme-catalyzed reaction can be developed so that these analogs are able to bind to enzyme active sites with comparable or even higher binding affinity than that of the native substrates, but are not able to be chemically transformed by enzyme active sites, the analogs would be able to compete favorably with the native substrates for binding to enzyme active sites, leading to the inhibition of the corresponding enzymatic reaction.

The inability of an enzyme active site to chemically transform a bound substrate analog could result from the lack of correctly positioned functional groups on the bound analog within the enzyme active site, so that no chemical reaction would be expected to occur. On the other hand, if the analog is a tighter binder than a native substrate, the corresponding Michaelis-like complex with the bound analog would be more stable (with lower free energy) than the Michaelis complex with the native substrate, and as such the activation barrier would be higher with the analog than that with the native substrate, making the chemical reaction with the analog more difficult than with the native substrate as far as only the reaction free energy profile is concerned.

The substrate-based inhibitors could be mono-substrate-, bi-substrate-, or multi-substrate-based analogs of the native substrate of an enzymatic reaction. The beauty with the bi-substrate and multi-substrate analog inhibitors is that selective inhibition among homologous enzymes catalyzing the same type of reaction would be potentially achievable with these compounds, which could result from the ability of a bi-substrate or multi-substrate analog inhibitor to bind with different affinities to different homologous enzymes that are able to catalyze the same type of reaction, yet have different three-dimensional topologies. Moreover, from the

Active Site-directed Enzyme Inhibitors: Design Concepts
By Weiping Zheng

Published by the Royal Society of Chemistry, www.rsc.org

thermodynamic consideration, a bi-substrate or multi-substrate analog inhibitor would exhibit a greater binding affinity than the corresponding mono-substrate analog inhibitor, because bi-valent or multi-valent binding of a bi-substrate or multi-substrate analog inhibitor, respectively, would engender a binding process with a more negative ΔG (free energy change) value than the mono-valent binding of a mono-substrate analog inhibitor due to the decreased entropic penalty with the binding of bi-substrate or multi-substrate analog inhibitors.

It should be noted that potent and selective bi-substrate and multi-substrate analog inhibitors can be developed only for the enzymatic reactions that obey sequential kinetic mechanism, in that the chemical step occurs only after all the substrates have been bound to the enzyme active site with the formation of the corresponding Michaelis complex.

3.2 Applications

As with the topic organization of Chapters 1 and 2, notable examples and the current status of development of the substrate-based inhibitors for each of the six types of the enzymatic reactions (*i.e.* the reactions catalyzed by transferases, oxidoreductases, hydrolases, lyases, isomerases, or ligases) are elaborated below. Given the above description and the fact that the working modes of the bi-substrate and multi-substrate analog inhibitors would be the same, only the bi-substrate analog inhibitors are covered to showcase the power of a substrate-based inhibitor composed of fragments derived from more than one substrate. The exception is mono-substrate analog inhibitors developed as chain terminators for the human immunodeficiency virus reverse transcriptase (HIV-RT)-catalyzed chain elongation reaction (an enzymatic transfer reaction), which hold potential as anti-HIV therapeutic agents.

3.2.1 Applications with Transferases

In this section, the mono-substrate analog inhibitors for the HIV-RT-catalyzed chain elongation reaction are elaborated first, followed by elaboration on notable examples of the bi-substrate analog inhibitors for various transferase-catalyzed reactions.

3.2.1.1 Mono-substrate Analog Inhibitors for the HIV-RT-catalyzed Chain Elongation Reaction

HIV-RT catalyzes the DNA chain elongation with the viral RNA as the template, and this enzymatic reaction is essential for the viral replication and amplification inside a host cell. Figure 3.1(A) depicts the reaction and the chemical mechanism for the HIV-RT-catalyzed DNA chain elongation reaction, whose driving force would be the pyrophosphatase-catalyzed hydrolysis of pyrophosphate to inorganic phosphate.[1] Given this mechanism, it would be straightforward and reasonable to speculate that the 2′,3′-dideoxy

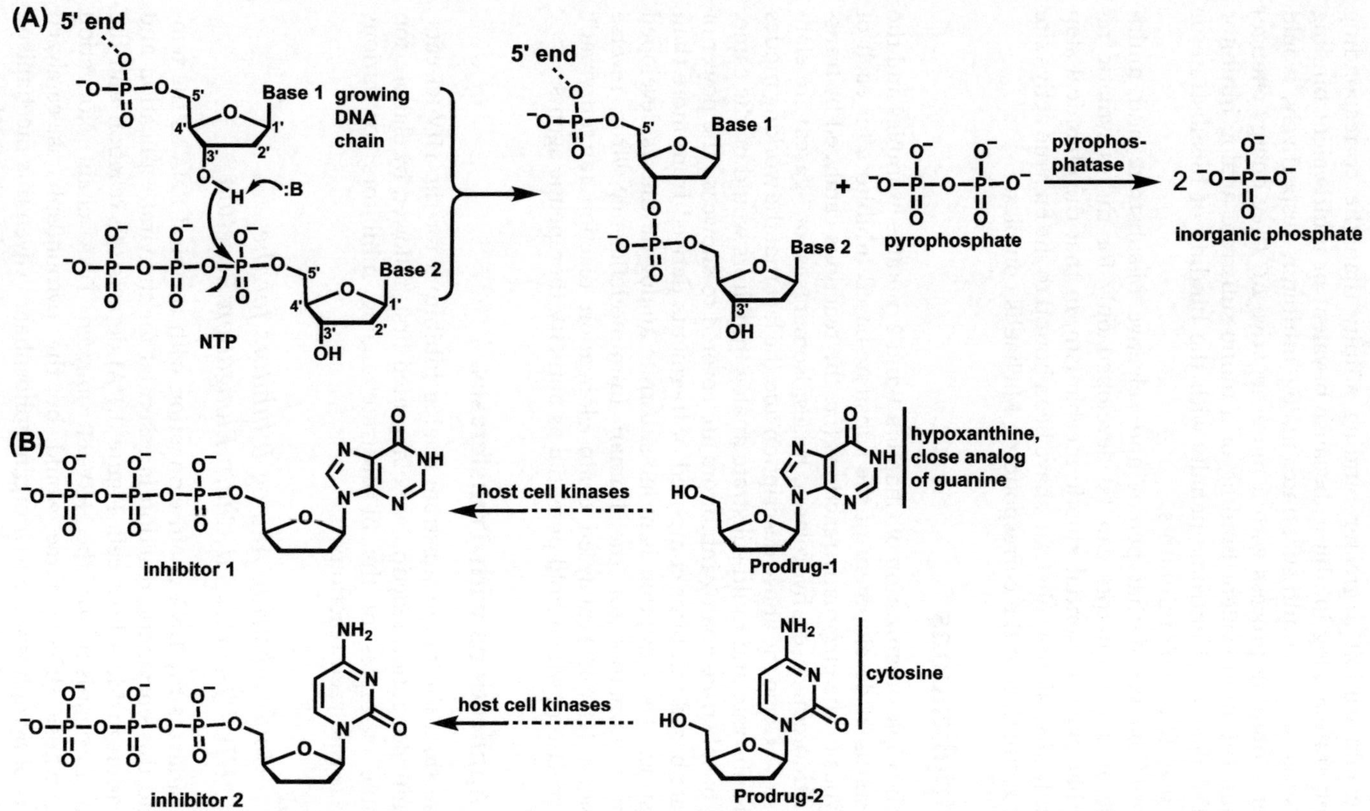

Figure 3.1 (A) The reaction and the chemical mechanism for the HIV-RT-catalyzed DNA chain elongation reaction. Bases 1 and 2, nucleobases found in a DNA strand; :B, a general base at the HIV-RT active site. (B) The chemical structures of the two representative mono-substrate 2′,3′-dideoxy analog inhibitors. Prodrug-1 and prodrug-2, the prodrug form of inhibitors 1 and 2, respectively.

analogs of a 2′-deoxy-nucleoside triphosphate (2′-deoxy-NTP) substrate could be taken by the enzyme as an alternate substrate, however, the chain elongation will be terminated after it has been incorporated into the growing DNA chain due to the lack of the 3′-hydroxy in its structure. Indeed, the two representative 2′,3′-dideoxy mono-substrate analogs depicted in Figure 3.1(B) have been found to be fairly potent inhibitors for the HIV-RT-catalyzed DNA chain elongation reaction.[1] Of note, in the structure of "inhibitor 1", the nucleobase was hypoxanthine (not a standard nucleobase found in DNA, but a close structural analog of guanine), which resulted from a structure–activity-relationship study. Due to the highly charged nature of "inhibitor 1" and "inhibitor 2", their respective nucleosides (*i.e.* "prodrug-1" and "prodrug-2" depicted in Figure 3.1(B)) have been also synthesized and fortuitously found to be converted inside a host cell to their respective triphosphate nucleotides (*i.e.* "inhibitor 1" and "inhibitor 2") by hijacking the host cell kinases, which have facilitated the development of the HIV-RT inhibitors as anti-HIV therapeutic agents.[1]

3.2.1.2 *A Bi-substrate Analog Inhibitor for the GCN5 Histone Acetyltransferase-catalyzed Acetylation Reaction*

A histone acetyltransferase (HAT) is able to catalyze the acetyl-coenzyme A (AcCoA)-dependent acetylation of specific lysine side chain ε-amino groups in both histone and non-histone proteins. The inhibition of the HAT-catalyzed acetylation reaction constitutes a potential therapeutic strategy for human diseases such as cancer. Figure 3.2(A) depicts the proposed chemical mechanism for the HAT-catalyzed acetylation reaction, which proceeds *via* the depicted tetrahedral intermediate.[2] It should be noted that the HAT-catalyzed acetylation obeys a sequential kinetic mechanism, which means that the two substrates (*i.e.* the protein lysine substrate and AcCoA) bind to a HAT to form a ternary complex before the first chemical step occurs. Given this mechanistic knowledge, the bi-substrate analog inhibitor depicted in Figure 3.2(B) was developed and found to be a fairly potent inhibitor for the HAT-catalyzed acetylation reaction.[2] As shown, this binary compound is a covalent conjugate of the two moieties derived from the two products (or substrates) of the HAT-catalyzed reaction.

3.2.1.3 *A Bi-substrate Analog Inhibitor for the Ghrelin O-Acyltransferase-catalyzed Octanoylation Reaction*

In addition to the above-described enzymatic acetylation reaction exemplified with the HAT-catalyzed AcCoA-dependent lysine N^{ε}-acetylation, enzymatic acylation (*i.e.* the installation of non-acetyl acyl groups) has also been reported, *e.g.* the ghrelin *O*-acyltransferase (GOAT)-catalyzed octanoyl-CoA-dependent serine *O*-octanoylation reaction shown in Figure 3.3(A).[3,4] Figure 3.3(A) also depicts the proposed chemical mechanism for the

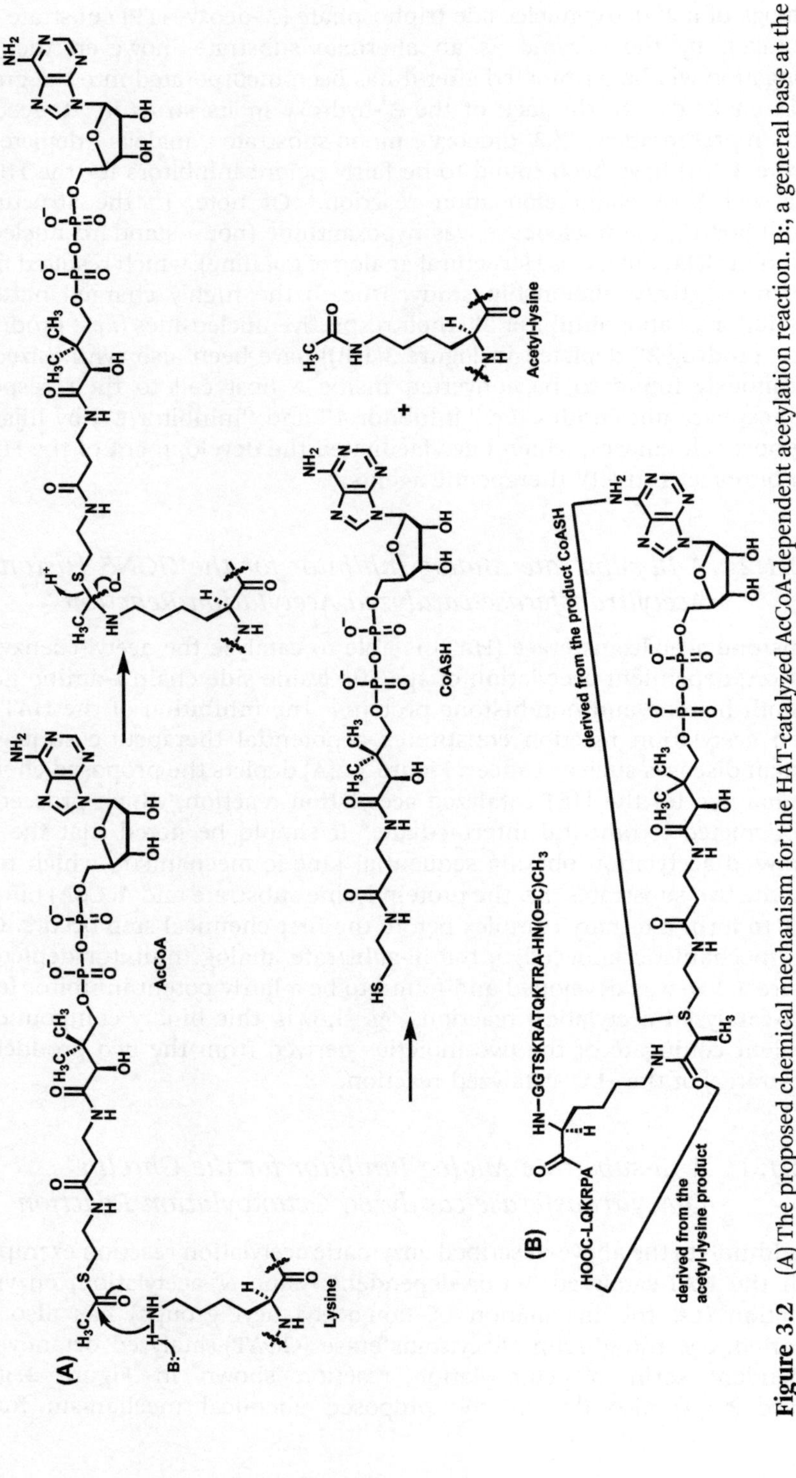

Figure 3.2 (A) The proposed chemical mechanism for the HAT-catalyzed AcCoA-dependent acetylation reaction. B:, a general base at the HAT active site. (B) The chemical structure of the designed bi-substrate analog inhibitor.

GOAT-catalyzed octanoylation reaction, which proceeds *via* the depicted tetrahedral intermediate, as depicted above for the HAT-catalyzed acetylation reaction. It should be noted that, like the HAT-catalyzed acetylation, the GOAT-catalyzed octanoylation presumably would also obey a sequential kinetic mechanism, which was supported by the experimentally observed fairly potent inhibition of the bi-substrate analog inhibitor depicted in Figure 3.3(B) against the GOAT-catalyzed octanoylation reaction. As another note, in addition to harboring a covalent conjugate of the two moieties respectively derived from the two products (or substrates) of the GOAT-catalyzed reaction as the key active component, this bi-substrate analog inhibitor also harbors a peptide sequence derived from the HIV protein Tat to enhance the cellular penetration.[3,4] Moreover, since the weight-promoting activity of the gastric peptide hormone ghrelin requires the GOAT-catalyzed octanoylation on the side chain of one of its serine residues, the inhibition of the GOAT-catalyzed octanoylation reaction would therefore constitute a potential therapeutic strategy for obesity and diabetes mellitus.[3,4]

In addition to the enzymatic transfer of acetyl and non-acetyl acyl groups as described above, other types of functional groups can also be transferred in various transferase-catalyzed reactions, as detailed below.

3.2.1.4 A Bi-substrate Analog Inhibitor for the Insulin Receptor Tyrosine Kinase-catalyzed Phosphorylation Reaction

Protein kinases (Tyr and Ser/Thr kinases) play crucial regulatory roles in cellular signaling by virtue of their ability to catalyze the transfer of the γ-phosphoryl group from ATP to a recipient hydroxyl group on the side chains of Tyr, Ser, or Thr, thereby changing the stereoelectronic properties of the side chain hydroxyl groups making them suitable to participate in the changing cellular signaling events. Insulin receptor tyrosine kinase is a typical Tyr kinase and has been shown to obey a sequential ternary complex kinetic mechanism in that the two substrates (*i.e.* the Tyr substrate and ATP) are able to bind to the active site of the enzyme before the chemical step occurs. Moreover, the phosphotransfer reaction catalyzed by insulin receptor tyrosine kinase was shown to proceed with a dissociative transition state, as shown in Figure 3.4(A).[5] Inspired by these mechanistic findings, the bi-substrate analog depicted in Figure 3.4(B) was designed and was found to be a fairly potent inhibitor against the reaction catalyzed by insulin receptor tyrosine kinase.[5] It should be noted that the two substrate-derived parts in this inhibitory compound were set apart with the depicted linker pursuant to the dissociative nature of the transition state of this enzymatic reaction. Moreover, the inhibitors of the insulin receptor tyrosine kinase could potentially be developed into therapeutic agents for human diseases such as cancer.

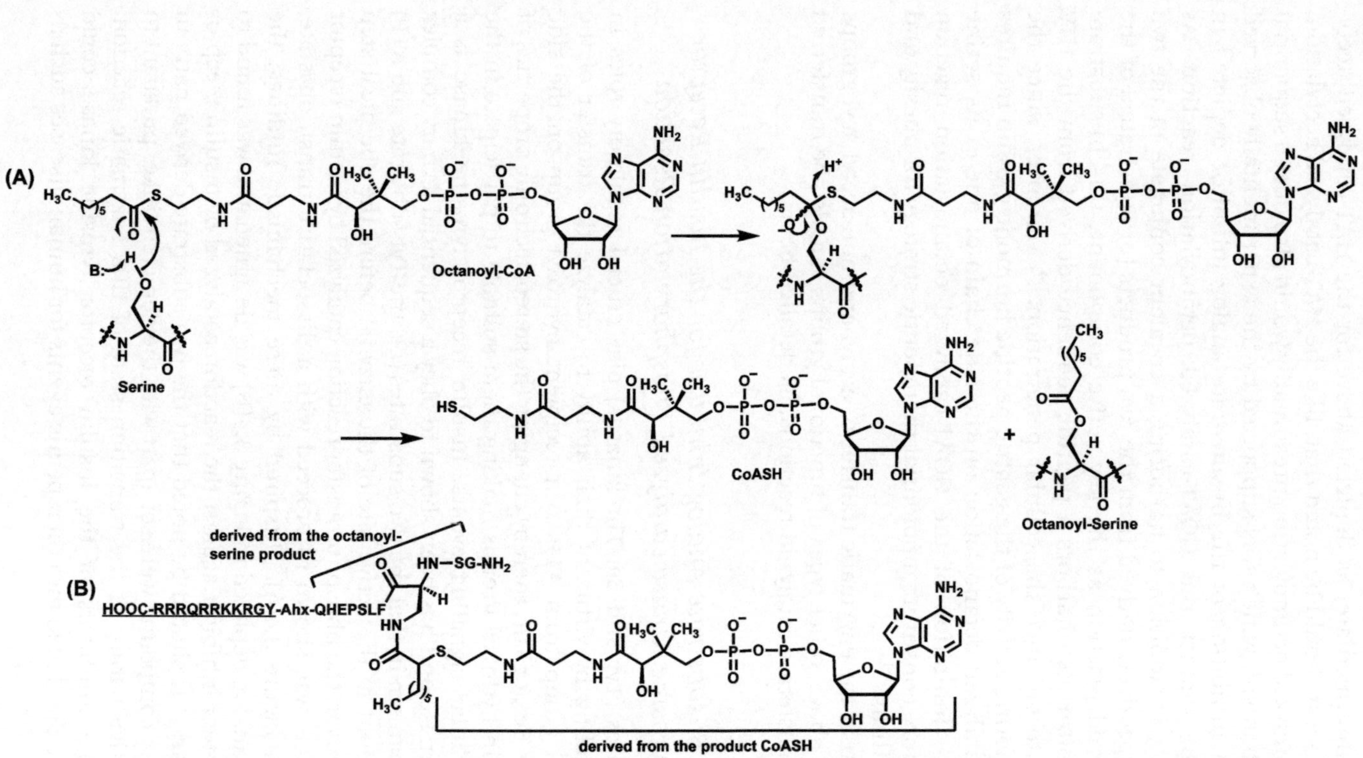

Figure 3.3 (A) The proposed chemical mechanism for the GOAT-catalyzed octanoyl-CoA-dependent octanoylation reaction. B:, a general base at the GOAT active site. (B) The chemical structure of the designed bi-substrate analog inhibitor. The underlined peptide sequence is that derived from the HIV protein Tat to enhance the cellular penetration. Ahx, 6-aminohexanoic acid.

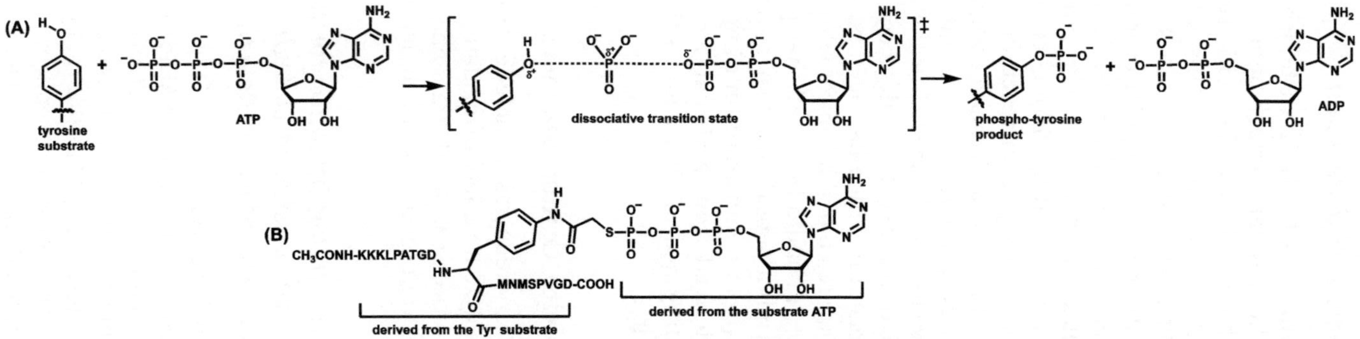

Figure 3.4 (A) The proposed chemical mechanism for the insulin receptor tyrosine kinase-catalyzed ATP-dependent phosphotransfer reaction. (B) The chemical structure of the designed bi-substrate analog inhibitor.

3.2.1.5 A Bi-substrate Analog Inhibitor for the O-GlcNAc Transferase-catalyzed Reaction

The mammalian enzyme *O*-GlcNAc transferase (OGT) catalyzes the *O*-GlcNAcylation (*i.e.* the covalent attachment of *N*-acetylglucosamine (GlcNAc)) on the side chain hydroxyl of specific serine or threonine residues on a cytosolic or nuclear protein. This enzymatic reaction uses uridine diphosphate (UDP)-GlcNAc as the GlcNAc donor, and plays an important role in regulating various cellular events such as transcription and cell cycle progression. Its inhibition holds a therapeutic potential against human diseases such as cancer and diabetes mellitus. Figure 3.5(A) depicts the proposed S_N2-like chemical mechanism for the OGT-catalyzed protein *O*-GlcNAcylation reaction.[6] The OGT-catalyzed reaction was also found to obey a ternary complex sequential kinetic mechanism in that both the protein substrate and UDP-GlcNAc need to be bound to the OGT active site before the chemical step occurs, and because of this, carefully designed bi-substrate analogs could be strong inhibitors for the OGT-catalyzed reaction, which has been borne out by the discovery of a fairly potent bi-substrate analog inhibitor whose chemical structure is shown in Figure 3.5(B).[7] It should be noted that part (a) in this chemical structure was derived from a peptide microarray screening to be a strong peptidic binder to the protein substrate binding site at OGT active site, whereas part (b) was derived from a virtual screening plus rational redesign as a strong binding neutral fragment to the UDP-GlcNAc binding site at the OGT active site.

3.2.2 Applications with Oxidoreductases

As mentioned above, oxidoreductases include oxidases, reductases, and dehydrogenases and catalyze a variety of oxidation or reduction reactions involved in various life processes. Bi-substrate analog inhibitors have been developed for several relevant reactions as detailed herein.

3.2.2.1 A Bi-substrate Analog Inhibitor for the Type 3 17β-Hydroxysteroid Dehydrogenase (Type 3 17β-HSD)-catalyzed Reaction

Type 3 17β-hydroxysteroid dehydrogenase (17β-HSD) catalyzes the nicotinamide adenine dinucleotide phosphate (NADPH)-dependent regioselective reduction of the C17 ketone of 4-androstene-3,17-dione (Δ^4-dione) to form the potent endogenous androgenic molecule testosterone, as depicted in Figure 3.6(A).[8] Therefore, the 17β-HSD-catalyzed reaction could be a therapeutic target for the androgen-dependent diseases, such as certain cancers. The binary compound depicted in Figure 3.6(B) was found to be a fairly potent bi-substrate analog inhibitor for the 17β-HSD-catalyzed

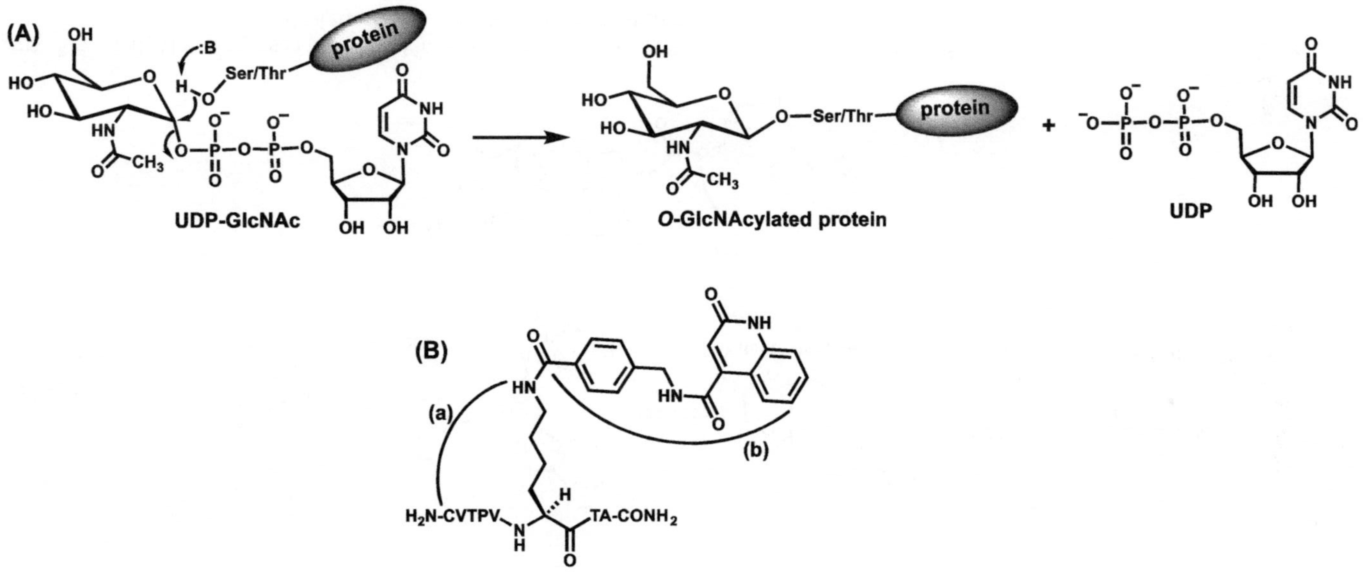

Figure 3.5 (A) The proposed chemical mechanism for the OGT-catalyzed protein *O*-GlcNAcylation reaction. :B, a general base at the OGT active site. (B) The chemical structure of the designed bi-substrate analog inhibitor.

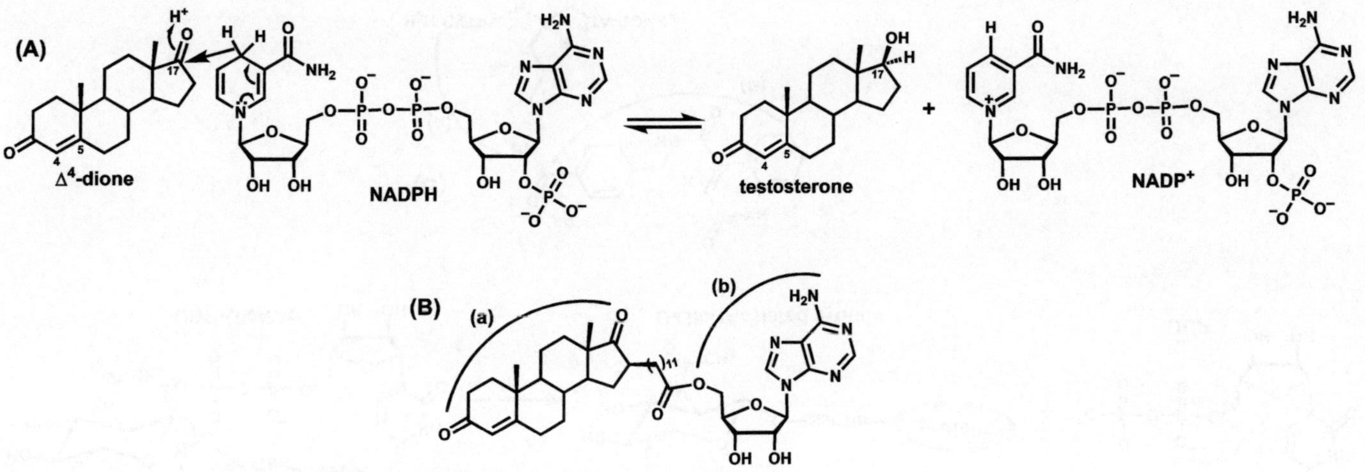

Figure 3.6 (A) The proposed chemical mechanism for the 17β-HSD-catalyzed regioselective reduction reaction. (B) The chemical structure of the designed bi-substrate analog inhibitor.

reaction, in which part (a) of the structure was derived from the reaction substrate Δ^4-dione and part (b) was derived from the reaction substrate NADPH.[8]

3.2.2.2 A Bi-substrate Analog Inhibitor for the Isocitrate Dehydrogenase-catalyzed Reaction

The isocitrate dehydrogenase (IDH)-catalyzed conversion of isocitrate to α-ketoglutarate, as shown in Figure 3.7(A), is an important metabolic reaction since it is the third reaction of the citric acid cycle, and its inhibitors would hold potential as research tools and therapeutic agents. The IDH-catalyzed reaction has a catalytic dependence on NAD^+ or $NADP^+$, as indicated in Figure 3.7(A).[9] As depicted, the first mechanistic step of the IDH-catalyzed reaction is the oxidation of the free hydroxyl of isocitrate to form oxalosuccinate whose β-ketocarboxylate moiety then undergoes spontaneous decarboxylation, affording the end product α-ketoglutarate. It should be noted that this chemical mechanistic scheme is consistent with a sequential ternary complex kinetic mechanism in that the two substrates (*i.e.* isocitrate and $NAD(P)^+$) are both bound to the IDH active site before the chemical step occurs, which would point to the feasibility of designing bi-substrate analog inhibitors for the IDH-catalyzed reaction. This has been borne out by the discovery of the fairly potent bi-substrate analog inhibitor depicted in Figure 3.7(B).[9]

3.2.2.3 A Bi-substrate Analog Inhibitor for the Dihydrofolate Reductase-catalyzed Reaction

Dihydrofolate reductase (DHFR) catalyzes the NADPH-dependent reduction of 7,8-dihydrofolate (DHF) to 5,6,7,8-tetrahydrofolate (THF), as depicted in Figure 3.8(A).[10] The further structural derivatization of THF would afford several THF derivatives that would behave as one-carbon donors in various biochemical reactions important in such cellular processes as the biosynthesis of the purine and pyrimidine precursors for nucleic acid biosynthesis. Therefore, the inhibitors for the DHFR-catalyzed reaction would ultimately lead to an inhibition of the biosynthesis of nucleic acids (*e.g.* DNA), which would be the molecular basis of the chemotherapeutic action of antifolate drugs. Figure 3.8(B) depicts the chemical structure of (2)-epigallocatechin-3-gallate (EGCG), a catechin derived from green tea, which has been found to be a fairly potent bi-substrate analog inhibitor for the DHFR-catalyzed reaction.[10] In the depicted chemical structure of EGCG, part (a) would be a structural mimic of the aromatic region of the DHFR reaction product THF and part (b) would mimic the nicotinamide moiety of $NADP^+$. Due to the structural similarity between DHF/NADPH and THF/$NADP^+$, this inhibitor is a *bona fide* bi-substrate analog.

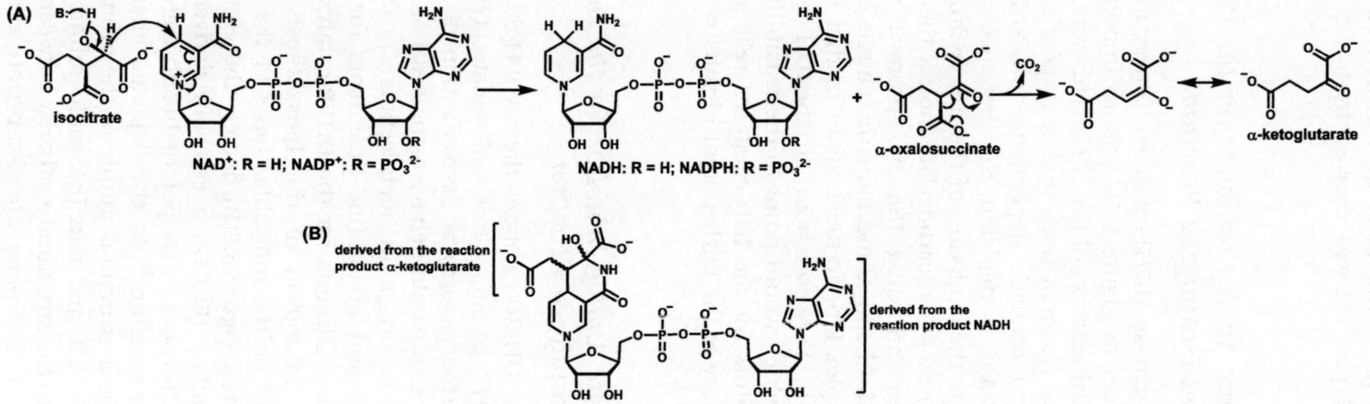

Figure 3.7 (A) The proposed chemical mechanism for the IDH-catalyzed reaction. B:, a general base at the IDH active site. (B) The chemical structure of the designed bi-substrate analog inhibitor, with the two structural moieties derived from the reaction product α-ketoglutarate and the reaction product NADH. Due to the structural similarity between α-ketoglutarate/NADH and isocitrate/NAD^+, this inhibitor is also a *bona fide* bi-substrate analog.

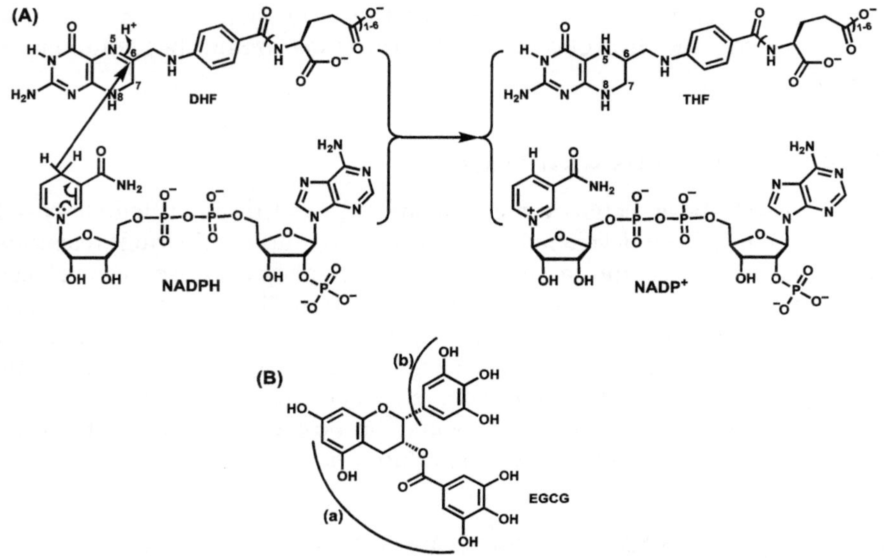

Figure 3.8 (A) The DHFR-catalyzed NADPH-dependent reduction of DHF to THF. (B) The chemical structure of the green tea catechin EGCG; part (a) would be a structural mimic of the aromatic region of THF and part (b) would mimic the nicotinamide moiety of $NADP^+$.

3.2.3 Applications with Hydrolases

As described above, the hydrolase-catalyzed reactions on both protein and non-protein substrates are involved in a variety of important life processes, and the substrates can be molecules harboring scissile amide (including β-lactam), acetal, ester (including phosphate, sulfate, and thioester), or thioether linkages. However, the bi-substrate and multi-substrate analog inhibitors have not yet been developed for hydrolase-catalyzed reactions, which could be due to the fact that all hydrolase-catalyzed reactions obey a ping-pong kinetic mechanism in that there is a formation of a covalent enzyme intermediate with the substrate to be hydrolyzed prior to the nucleophilic attack of another substrate (*i.e.* water) onto the covalent enzyme intermediate. Nevertheless, mono-substrate analog inhibitors have been developed for hydrolase-catalyzed reactions, yet a "pure" or "naive" mono-substrate analog inhibitor such as an oligo-peptide simply and directly derived from a substrate protein at most would exhibit a modest inhibitory potency for the cognate hydrolase-catalyzed reaction. One remedy for such scenario would be and has been to judiciously equip the mono-substrate analog inhibitors with modest inhibitory potencies with certain inhibitory "warheads", so as to transform them into the catalytic mechanism – or the catalytic intermediate-based inhibitors described above, or the transition state analog inhibitors or the (photo)affinity labeling-/covalent-type inhibitors described below. In view of this, the mono-substrate analog inhibitors for the hydrolase-catalyzed reactions would be redundantly described here.

3.2.4 Applications with Lyases

As discussed above, catalytic mechanism-and catalytic intermediate-based inhibitors have been developed for the lyation (group elimination) reactions catalyzed by lyases and other enzymes (*e.g.* decarboxylase) which play important regulatory roles in a variety of important life processes. However, the substrate-based inhibitors including mono-substrate, bi-substrate, and multi-substrate analog inhibitors have not yet been developed for enzymatic lyation reactions. Given that certain enzymatic lyation reactions obey a sequential kinetic mechanism, it would be possible to develop their bi-substrate and multi-substrate analog inhibitors.

3.2.5 Applications with Isomerases: A Bi-substrate Analog Inhibitor for the 1-Deoxy-D-xylulose 5-Phosphate Reductoisomerase (MEP Synthase)-catalyzed Reaction

As described above, isomerases catalyze a variety of reactions involved in various important life processes, and the catalytic mechanism-and intermediate-based inhibitors have been developed for the isomerase-catalyzed reactions and for the reaction catalyzed by a racemase. It is exciting to see that a fairly potent bi-substrate analog inhibitor has also been developed for an enzymatic isomerization reaction, *i.e.* that catalyzed by 1-deoxy-D-xylulose 5-phosphate reductoisomerase (MEP synthase), which is the first committed step for the biosynthesis of isopentenyl pyrophosphate (IPP) and dimethylallyl pyrophosphate (DMAPP), the two building blocks for the ultimate synthesis of the isoprenoid family of crucial biomolecules. Since mammals, including humans, use a different MEP pathway to synthesize IPP and DMAPP than certain bacteria, the inhibitors for the bacterial MEP synthase-catalyzed reaction could potentially be developed into novel antibiotics.

Figure 3.9(A) depicts the MEP synthase-catalyzed manganese (Mn^{2+})-dependent reaction.[11] As depicted, the initial isomerization of the substrate 1-deoxy-D-xylulose 5-phosphate (DXP) to the depicted intermediate is followed by the NADPH-dependent reduction of the aldehydic functionality in the intermediate, affording the product of this enzymatic reaction, *i.e.* 2-*C*-methyl-D-erythritol 4-phosphate (MEP).

Given that the MEP synthase-catalyzed reaction obeys a sequential kinetic mechanism, it seems to be feasible to develop bi-substrate analog inhibitors for this bi-substrate enzymatic reaction. Indeed, the compound depicted in Figure 3.9(B) was found to be a fairly potent inhibitor for the MEP synthase-catalyzed reaction.[12] Of note, part (a) of the compound is a close structural analog of the substrate DXP, with the two oxygen atoms of the hydroxamate

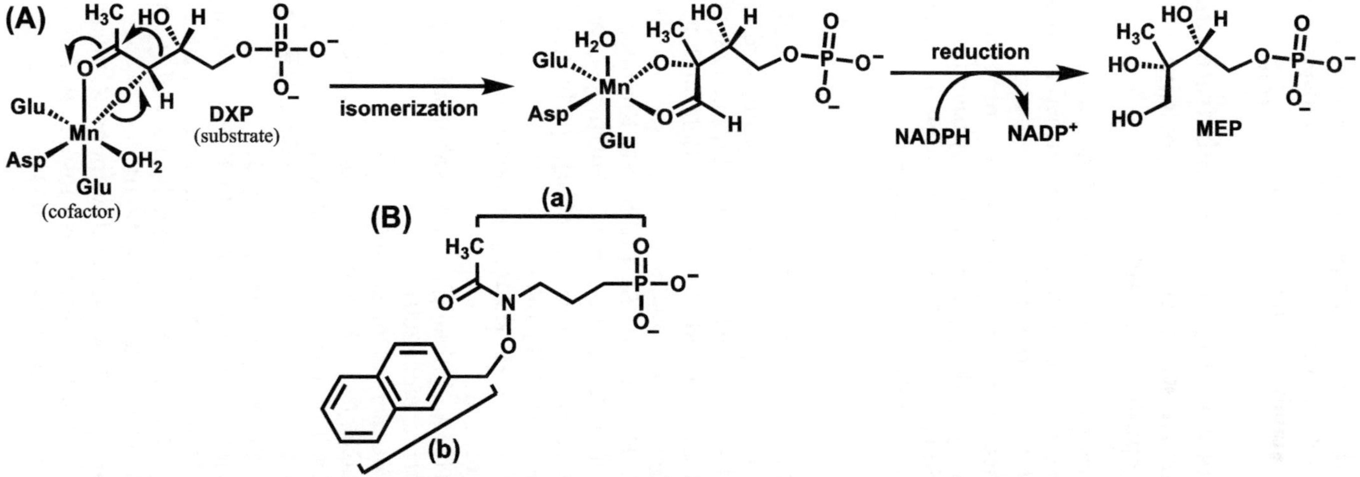

Figure 3.9 (A) The proposed chemical mechanism for the MEP synthase-catalyzed manganese (Mn^{2+})-dependent reaction. Mn, Mn^{2+}; DXP, 1-deoxy-D-xylulose 5-phosphate; MEP, 2-*C*-methyl-D-erythritol 4-phosphate. Please see other figures (*e.g.* Figure 3.7) in this book for the full chemical structures of $NADP^+$ and NADPH. (B) The chemical structure of the designed bi-substrate analog inhibitor, with part (a) being a close structural analog of DXP; part (b) would occupy the NADPH binding pocket on the enzyme.

moiety being also able to chelate with Mn^{2+}; part (b) of the molecule would be able to occupy the NADPH binding pocket on the enzyme. Moreover, the use of phosphonate in the compound instead of phosphate in DXP would enhance the chemical stability of the compound.

3.2.6 Applications with Ligases: A Bi-substrate Analog Inhibitor for the Glutathionylspermidine Synthetase-catalyzed Reaction

As stated above, ligases catalyze a variety of ligation reactions involved in important life processes. Several catalytic intermediate-based inhibitors have been developed for various ligase-catalyzed reactions, for example the catalytic intermediate-based phosphinate inhibitor described in Section 2.2.6.3 for the glutathionylspermidine synthetase (GspS)-catalyzed reaction. As depicted in Figure 2.19, this GspS inhibitor is also a bi-substrate analog inhibitor for the GspS-catalyzed reaction.

References

1. R. A. Copeland, R. R. Gontarek and L. Luo, in *Textbook of drug design and discovery*, ed. P. Krogsgaard-Larsen, K. Stromgaard and U. Madsen, CRC Press, Boca Raton, 4th edn, 2010, ch. 11, pp. 178–179.
2. A. N. Poux, M. Cebrat, C. M. Kim, P. A. Cole and R. Marmorstein, *Proc. Natl. Acad. Sci. U.S.A.*, 2002, **99**, 14065.
3. B. P. Barnett, Y. Hwang, M. S. Taylor, H. Kirchner, P. T. Pfluger, V. Bernard, Y.-Y. Lin, E. M. Bowers, C. Mukherjee, W.-J. Song, P. A. Longo, D. J. Leahy, M. A. Hussain, M. H. Tschöp, J. D. Boeke and P. A. Cole, *Science*, 2010, **330**, 1689.
4. M. S. Taylor, Y. Hwang, P.-Y. Hsiao, J. D. Boeke and P. A. Cole, *Methods Enzymol.*, 2012, **514**, 205.
5. K. Parang, J. H. Till, A. J. Ablooglu, R. A. Kohanski, S. R. Hubbard and P. A. Cole, *Nat. Struct. Biol.*, 2001, **8**, 37.
6. E. J. Kim, *ChemBioChem*, 2020, **21**, 3026.
7. H. Zhang, T. Tomašič, J. Shi, M. Weiss, R. Ruijtenbeek, M. Anderluh and R. J. Pieters, *Med. Chem. Commun.*, 2018, **9**, 883.
8. M. Bérubé and D. Poirier, *J. Enzyme Inhib. Med. Chem.*, 2007, **22**, 201.
9. R. S. Ehrlich, *J. Enzyme Inhib.*, 2000, **15**, 265.
10. M. Spina, M. Cuccioloni, M. Mozzicafreddo, F. Montecchia, S. Pucciarelli, A. M. Eleuteri, E. Fioretti and M. Angeletti, *Proteins*, 2008, 72, 240.
11. S. Steinbacher, J. Kaiser, W. Eisenreich, R. Huber, A. Bacher and F. Rohdich, *J. Biol. Chem.*, 2003, **278**, 18401.
12. A. Haymond, C. Johny, T. Dowdy, B. Schweibenz, K. Villarroel, R. Young, C. J. Mantooth, T. Patel, J. Bases, G. San Jose, E. R. Jackson, C. S. Dowd and R. D. Couch, *PLoS One*, 2014, **9**, e106243.

CHAPTER 4

Transition State-based Design

4.1 Mode of Working

As a consensus in the field of enzymology, the underlying mechanism by which the majority of enzymes accelerate chemical reactions is to preferentially stabilize the transition state of a chemical reaction over the ground state substrate, thereby diminishing the activation free energy ($\Delta G^{\ddagger}$), leading to rate enhancement. If chemically stable close structural analogs of a transition state could be developed such that the enzyme would bind and preferentially stabilize the analogs over the ground state substrate, yet be unable to chemically transform them, these analogs would then be able to prevent the enzyme from binding and processing the normal substrate, thereby leading to inhibition of the enzymatic reaction.

4.2 Applications

As with the topic organization of Chapters 1–3, notable examples and the current status of development of the transition state analog inhibitors for each of the six types of the enzymatic reactions (*i.e.* the reactions catalyzed by transferases, oxidoreductases, hydrolases, lyases, isomerases, or ligases) are elaborated here.

4.2.1 Applications with Transferases

In this section, notable examples of the transition state analog inhibitors for various transferase-catalyzed reactions are elaborated.

Active Site-directed Enzyme Inhibitors: Design Concepts
By Weiping Zheng

Published by the Royal Society of Chemistry, www.rsc.org

4.2.1.1 A Transition State Analog Inhibitor for the Glycosyltransferase MurG-catalyzed Reaction

The *Escherichia coli* MurG is a glycosyltransferase able to catalyze the transfer of *N*-acetyl-D-glucosamine (GlcNAc) from uridine diphosphate (UDP)-GlcNAc to the free 4′-hydroxyl on lipid intermediate I, giving rise to lipid intermediate II, as shown in Figure 4.1(A).[1] The lipid intermediate II can be further transformed ultimately to the peptidoglycan structure in the bacterial cell wall. The inhibition of the MurG-catalyzed reaction has been regarded as a strategy for developing novel antibacterial agents. Figure 4.1(A) also depicts the proposed chemical mechanism for the MurG-catalyzed reaction.[1] One salient feature of this mechanism is the formation of the depicted putative transition state structure bestowed with an intra-ring oxonium cation, which could be profitably exploited in inhibitor design. Indeed, the transition state analog inhibitor depicted in Figure 4.1(B) was designed to take advantage of exactly this structural feature and was found to behave as a reasonably potent inhibitor for the MurG-catalyzed reaction.[1] In this inhibitor, the intra-ring protonated amine (under physiological pH) would mimic the intra-ring oxonium cation in the putative transition state structure. Moreover, since a co-crystal structure of MurG in complex with UDP-MurNAc (a close structural analog of UDP-GlcNAc) revealed that the uridine moiety constituted an important binding determinant with MurG, and there were no electrostatic interactions between the negatively charged diphosphate bridge (under physiological pH) and the enzyme, therefore, in the designed inhibitor, uridine was included and a neutral bridge was used to replace the diphosphate bridge, as shown in Figure 4.1(B). Further, per the crystal structure, the substituted bicyclic ring structure in the designed inhibitor could be also favorably bound at MurG active site.

4.2.1.2 A Transition State Analog Inhibitor for the Spermine/Spermidine-N(1)-acetyltransferase-catalyzed Reaction

The spermine/spermidine-*N(1)*-acetyltransferase (SSAT) catalyzes the rate-limiting step in the catabolism of the polyamines including spermine and spermidine inside mammalian cells. The inhibitors of this reaction would be useful tools in helping to further decipher the polyamine biology. Figure 4.2(A) depicts the proposed chemical mechanism for the SSAT-catalyzed acetyl-coenzyme A (AcCoA)-dependent acetylation reaction.[2] The key feature of this mechanism is the formation of a transition state with the depicted putative chemical structure. Since the SSAT-catalyzed reaction appears to obey a sequential ternary complex kinetic mechanism, it would be feasible to design a covalently conjugated binary compound with two constituent moieties respectively derived from the two substrates of this enzymatic reaction (*i.e.* spermine (or spermidine) and AcCoA) or the two products of the reaction (*i.e.* *N(1)*-acetyl-spermine (or *N(1)*-acetyl-spermidine) and coenzyme A (CoASH)), while incorporating a transition state structural mimic.

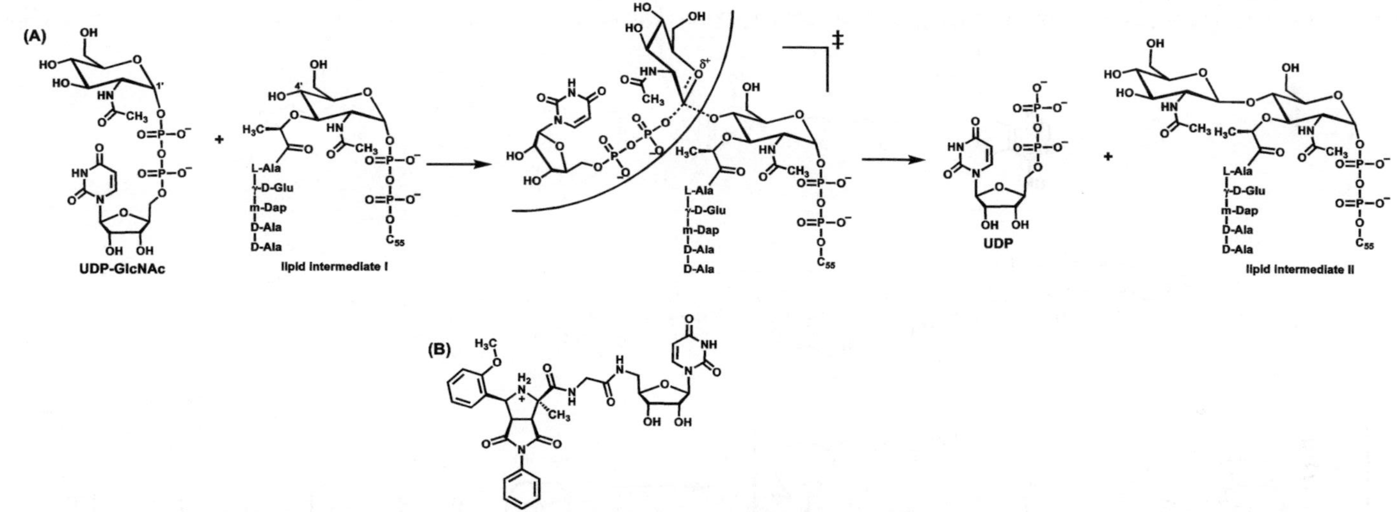

Figure 4.1 (A) The proposed chemical mechanism for the MurG-catalyzed reaction. C_{55}, undecaprenyl. The half-circled portion of the putative transition state structure is that primarily mimicked in designing transition state analog inhibitors. (B) The chemical structure of a transition state analog inhibitor of the MurG-catalyzed reaction, which would structurally mimic the transition state formed during the normal enzymatic catalysis as depicted in (A). The protonated intra-ring amine is supposed to mimic the intra-ring oxonium cation in the putative transition state structure.

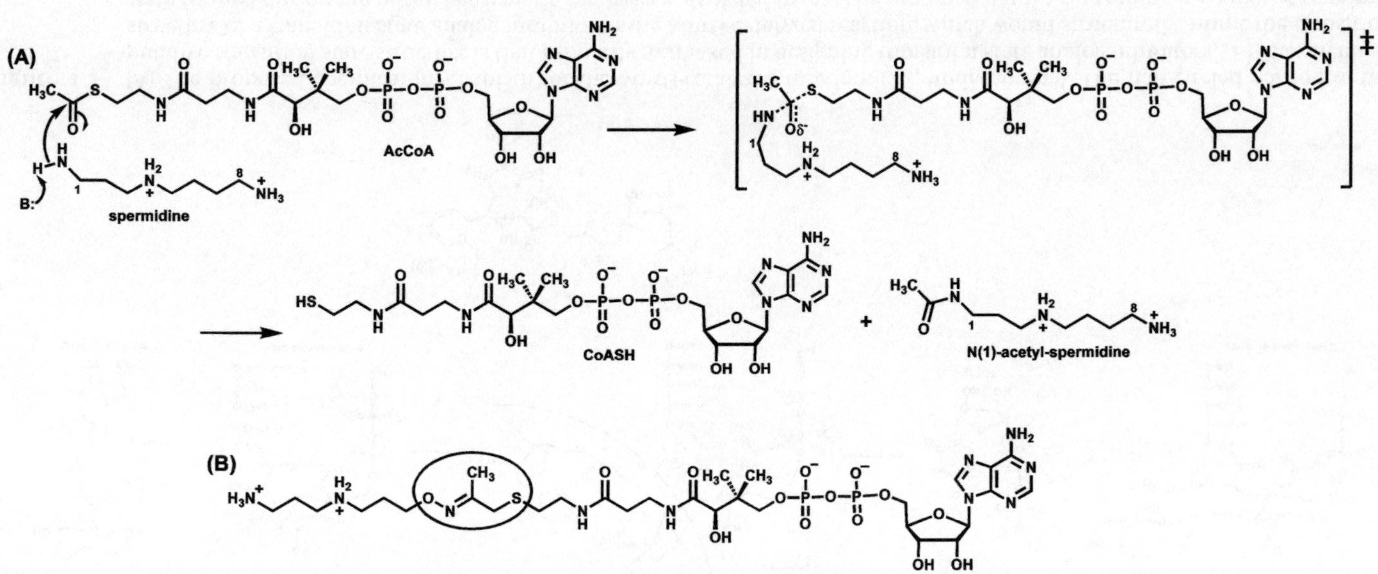

(A) spermidine AcCoA CoASH N(1)-acetyl-spermidine (B)

Figure 4.2 (A) The proposed chemical mechanism for the SSAT-catalyzed reaction, as illustrated with that on the substrate spermidine. B:, a general base at the SSAT active site. (B) The chemical structure of a transition state analog inhibitor of the SSAT-catalyzed reaction, with the transition state structural mimic circled.

Figure 4.2(B) depicts one such inhibitor designed as a covalently conjugated binary compound with two constituent moieties derived from the two products of the reaction (*i.e. N(1)*-acetyl-spermidine and CoASH), with the transition state structural mimic circled.

4.2.1.3 *A Transition State Analog Inhibitor for Ribosomal Peptidyl Transferase-catalyzed Peptide Bond Formation*

Ribosomal peptidyl transferase activity catalyzes the peptide bond formation between the C-terminal carbonyl of a polypeptide chain loaded on transfer RNA (tRNA) at the P site of the ribosome and the α-amino of an amino acid loaded on tRNA at the A site of the ribosome, thus leading to the elongation of the polypeptide chain. Figure 4.3(A) depicts the proposed chemical mechanism of the peptidyl transferase-catalyzed peptide bond formation on ribosome.[3] The key feature of this mechanism is the formation of the depicted putative transition state structure. Given the potential of the inhibitors for this enzymatic reaction as novel antibiotics and tools for enriching and characterizing this enzymatic activity, transition state analog inhibitors have been pursued with one such designed inhibitor depicted in Figure 4.3(B).[3] Salient structural features of this inhibitor include the use of phosphoramidate (circled portion) as a mimic of the circled transition state structural moiety depicted in Figure 4.3(A) and the inclusion of the 5′-nonphosphorylated 2′-deoxyadenosine analog of the 3′-terminal trinucleotide CCA ubiquitously present in tRNA (*i.e.* CC(dA)) and the antibiotic puromycin able to bind to the A site on the ribosome. In addition to mimicking the key transition state structural moiety (circled portion in Figure 4.3(A)), the logic behind constructing such a binary inhibitory compound is that the two substrates of the ribosomal peptidyl transferase activity, at P site and A site of ribosome, seemingly need to be aligned into a proper relative geometry before a direct peptidyl transfer occurs.

4.2.2 Applications with Oxidoreductases

In this section, notable examples of the transition state analog inhibitors for several oxidoreductase-catalyzed reactions are elaborated.

4.2.2.1 *A Transition State Analog Inhibitor for the 6-Phosphogluconate Dehydrogenase-catalyzed Reaction*

As depicted in Figure 2.6(A) and Figure 4.4(A), the pentose phosphate shunt enzyme 6-phosphogluconate dehydrogenase (6PGDH) catalyzes the conversion of 6-phosphogluconate (6PG) to ribulose-5-phosphate (Ru5P) concomitant with the reduction of $NADP^+$ to the reducing equivalent NADPH. As depicted in both figures, the 6PGDH-catalyzed reaction is composed of two stages: dehydrogenation and decarboxylation. The proposed transition

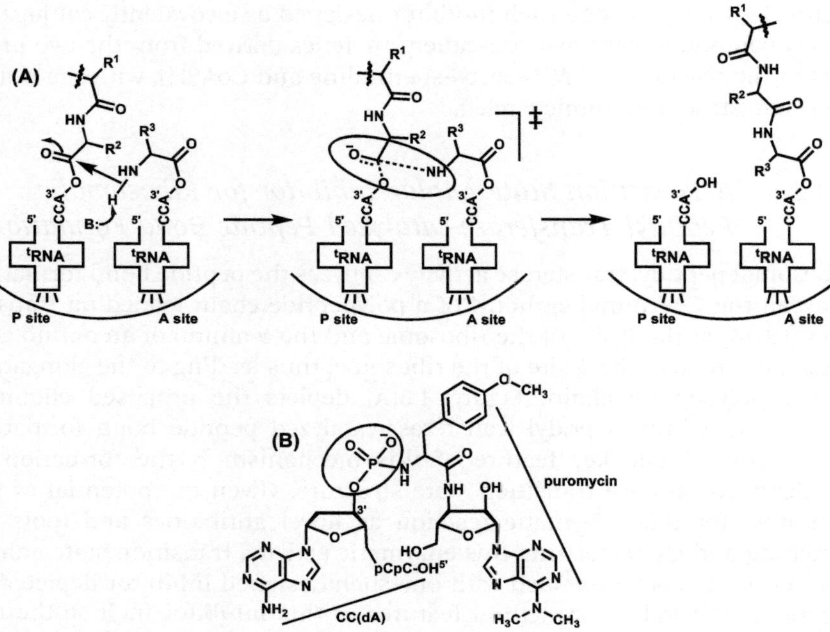

Figure 4.3 (A) The proposed chemical mechanism for the ribosomal peptidyl transferase activity-catalyzed peptide bond formation on the P (peptidyl) and A (amino acid) sites on ribosome. CCA, the 3′-terminal trinucleotide ubiquitously present in tRNA; $R^1/R^2/R^3$, amino acid side chains. The key transition state structural moiety is circled. As one reaction product, the unloaded tRNA will be translocated to the E (exit) site (not shown) for exit from the ribosome; as another reaction product, the elongated polypeptide chain (by one amino acid residue) will be translocated to the P site for further chain elongation; the empty A site will then be loaded with another aminoacyl-tRNA molecule. (B) The chemical structure of a transition state analog inhibitor of ribosomal peptidyl transferase activity-catalyzed peptide bond formation, with the transition state structural mimic circled. The structural moiety on the left of this structural mimic is the 5′-nonphosphorylated 2′-deoxyadenosine analog of CCA trinucleotide (*i.e.* CC(dA)) and that on the right is the antibiotic puromycin able to bind to the A site on the ribosome.

state structures for the dehydrogenation stage are indicated in Figure 4.4(A), with the key structural moieties circled.

While the high-energy 1,2-enediolate intermediate depicted in Figure 2.6(A) and Figure 4.4(A) formed during the decarboxylation stage of the 6PGDH-catalyzed reaction has been exploited as the basis for the design of a potent catalytic intermediate-based inhibitor for the 6PGDH-catalyzed reaction (described in Section 2.2.2.3 and depicted in Figure 2.6(B)), a potent transition state analog inhibitor (shown in Figure 4.4(B)) for the 6PGDH-catalyzed reaction has also been designed to structurally mimic the "late" transition state depicted in Figure 4.4(A) for the 6PGDH-catalyzed dehydrogenation reaction.[4]

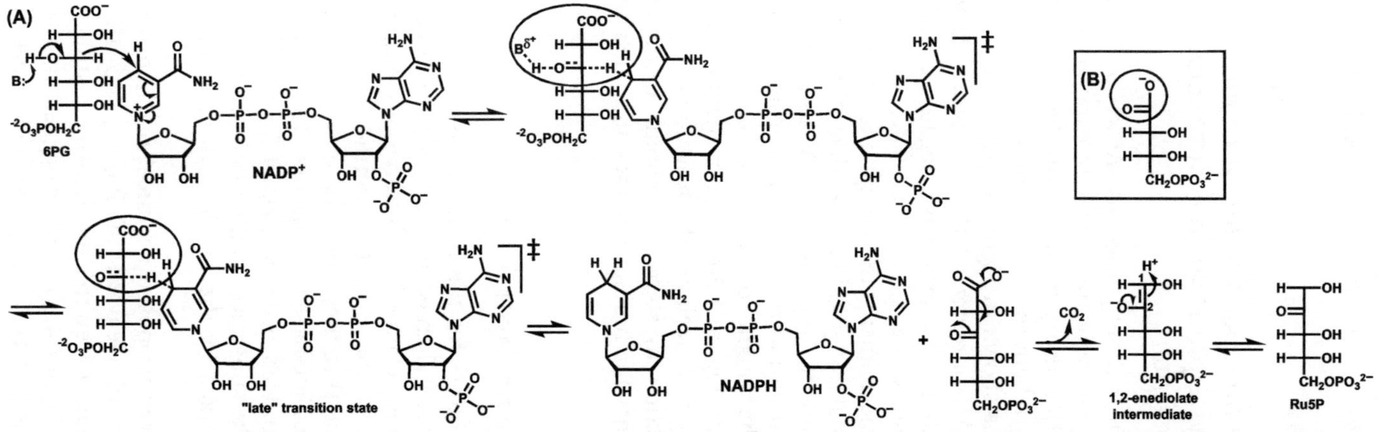

Figure 4.4 (A) The proposed chemical mechanism for the 6PGDH-catalyzed reaction which is composed of the dehydrogenation and decarboxylation stages. The putative transition state structures for the dehydrogenation stage are indicated, with the key structural moieties circled. B:, a general base at 6PGDH active site. (B) The chemical structure of the transition state analog inhibitor for the dehydrogenation stage and thus the entire 6PGDH-catalyzed reaction, with the circled moiety mimicking that circled in the chemical structure of the "late" transition state depicted in (A). Note: some of the mechanistic details are also depicted in Figure 2.6.

Of note, the free carboxylate anion of the inhibitor is the key moiety mimicking the circled anionic moiety in the chemical structure of the "late" transition state.

4.2.2.2 *A Transition State Analog Inhibitor for the Type 5 17β-Hydroxysteroid Dehydrogenase-catalyzed Reductive Reaction*

The type 5 17β-hydroxysteroid dehydrogenase AKR1C3 belongs to the aldo-keto reductase (AKR) superfamily and is able to catalyze the oxidoreductive reactions on different 17β-hydroxysteroid substrates, such as the reduction of the weak androgen Δ^4-androstene-3,17-dione to furnish the potent androgen testosterone, as shown in Figure 4.5(A). As depicted, the NADPH-dependent reduction of Δ^4-androstene-3,17-dione proceeds *via* the indicated oxyanionic transition state in which there is a negative charge buildup about the C17 position of the substrate.[5]

Given this chemical mechanism, the compound shown in Figure 4.5(B) was designed as a potential transition state analog inhibitor for the AKR1C3-catalyzed NADPH-dependent reduction of Δ^4-androstene-3,17-dione to testosterone.[5] The key structural moiety in this designed compound is the 17β-carboxylate anion that ought to mimic the oxyanion in the transition state structure depicted in Figure 4.5(A). The availability of such an inhibitory compound would help to decipher the functional roles of the AKR1C3-catalyzed reductive reaction in steroid hormone action and could even furnish therapeutic agents for human diseases in which there is a dysregulated hormonal action.

4.2.3 Applications with Hydrolases

Even though many inhibitors for the hydrolase-catalyzed reactions on the amide substrates have been claimed in the literature to be transition-state analog inhibitors, all these claimed inhibitory compounds were found to actually structurally mimic the catalytic intermediate instead of the preceding transition state, so that they would be more appropriately classified as the catalytic intermediate-based inhibitors. However, transition state analog inhibitors have indeed been developed for the hydrolase-catalyzed reaction on the acetal-like substrate, as illustrated below with the *N*-ribosyl hydrolysis reaction catalyzed by the inosine-uridine preferring nucleoside hydrolase (IU-NH).[6]

Figure 4.6(A) depicts the proposed chemical mechanism for the IU-NH-catalyzed *N*-ribosidic bond cleavage of inosine *via* the proposed oxocarbenium transition state. Given this mechanistic feature, the iminoribitol compound depicted in Figure 4.6(B) was designed in the hope that the protonated iminoribitol ring NH under physiological pH would mimic the oxocarbenium cation in the proposed transition state depicted in

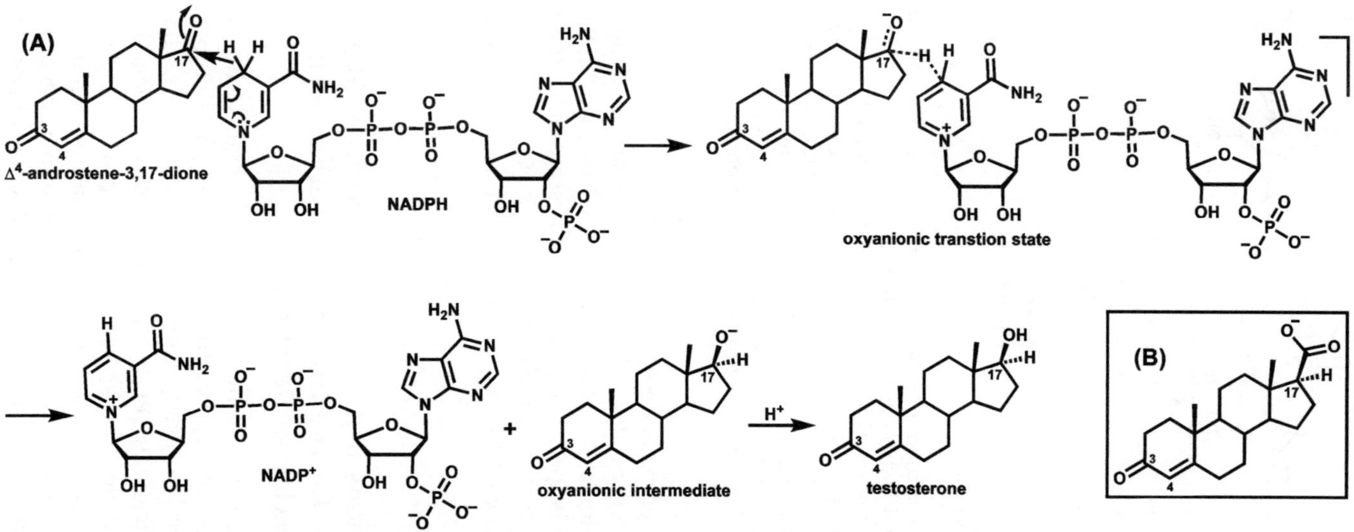

Figure 4.5 (A) The proposed chemical mechanism for the AKR1C3-catalyzed NADPH-dependent reduction of Δ^4-androstene-3,17-dione to testosterone. The proposed oxyanionic transition state structure is indicated. (B) The chemical structure of a potential transition state analog inhibitor for the AKR1C3-catalyzed reductive reaction, with the 17β-carboxylate anion in the structure mimicking the oxyanion in the chemical structure of the transition state depicted in (A).

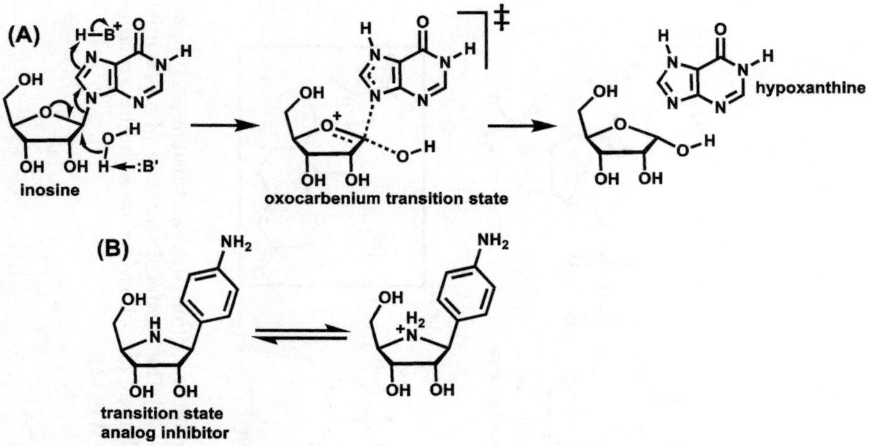

Figure 4.6 (A) The proposed chemical mechanism for the IU-NH-catalyzed *N*-ribosyl hydrolysis (*i.e.* *N*-ribosidic bond cleavage) of inosine to give rise to hypoxanthine and the aglycone product. The proposed oxocarbenium transition state structure is indicated. (B) The chemical structure of a transition state analog inhibitor for the IU-NH-catalyzed hydrolytic reaction, with the protonated iminoribitol ring NH in the structure under physiological pH mimicking the oxocarbenium cation in the chemical structure of the transition state depicted in (A).

Figure 4.6(A). Indeed, this compound was found to be a fairly potent inhibitor for the IU-NH-catalyzed hydrolytic reaction, implying that it would behave as a transition state analog inhibitor. Given that the IU-NH-catalyzed hydrolytic reaction is specifically present and essential for the purine salvage in trypanosomal parasites which lack the *de novo* purine biosynthetic capacity, its inhibitors could be developed into anti-trypanosomal therapeutic agents.

4.2.4 Applications with Lyases: A Transition State Analog Inhibitor for the Ornithine Decarboxylase-catalyzed Decarboxylation Reaction

In addition to the catalytic mechanism- and catalytic intermediate-based inhibitors described above, transition state analog inhibitors have also been developed for a lyation (group elimination) reaction, *i.e.* that catalyzed by the pyridoxal 5′-phosphate (PLP)-dependent ornithine decarboxylase (ODC). The ODC-catalyzed decarboxylation reaction produces putrescine, as depicted in Figure 4.7(A), from where other polyamines can be biosynthesized. Therefore, this enzymatic reaction plays an important regulatory role in both prokaryotic and eukaryotic physiology. The inhibitors of this enzymatic reaction would possess anti-tumor and anti-parasitic potential.

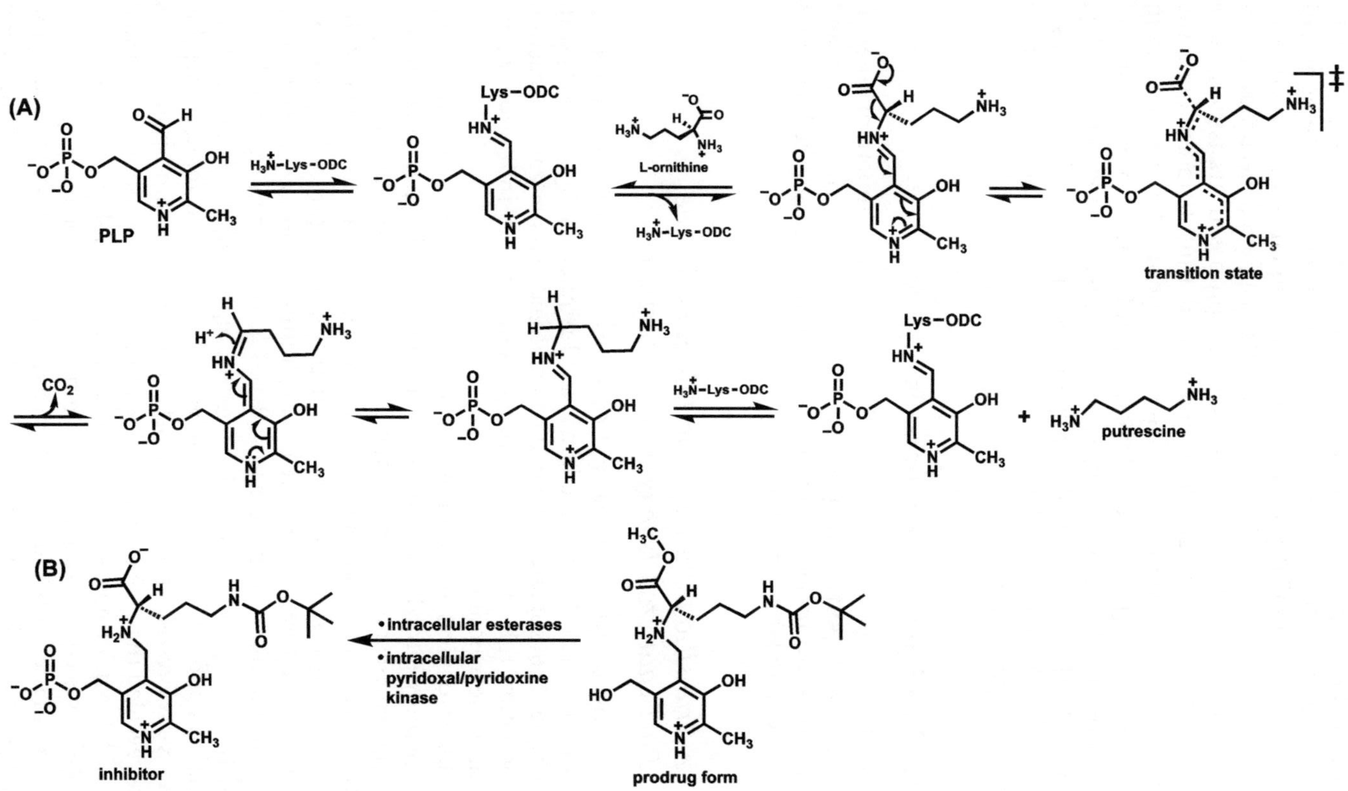

Figure 4.7 (A) The proposed chemical mechanism for the ODC-catalyzed PLP-dependent deacrboxylation reaction. The proposed transition state structure is indicated. (B) The chemical structure of a transition state analog inhibitor for the ODC-catalyzed decarboxylation reaction and its more cell-permeable prodrug form which can be enzymatically hydrolyzed and phosphorylated to the active inhibitor once inside cells.

Figure 4.7(A) depicts the proposed chemical mechanism for the ODC-catalyzed PLP-dependent decarboxylation reaction.[7,8] One salient mechanistic feature is the extensive electron delocalization in the proposed transition state structure. Figure 4.7(B) depicts the chemical structure of a transition state analog inhibitor for the ODC-catalyzed decarboxylation reaction, which was designed to mimic the proposed transition state structure. To eliminate the negative charge and thus help enhance the bioavailability of this inhibitor, its more cell permeable prodrug form (depicted in Figure 4.7(B)) was developed, and was found to be capable of potently inhibiting the ODC catalytic activity inside cells. Presumably, once inside cells, the prodrug could be hydrolyzed by intracellular esterases and regioselectively phosphorylated by pyridoxal/pyridoxine kinase, thus furnishing the active inhibitor *in situ*.[7–9]

4.2.5 Applications with Isomerases

As stated above, the isomerase-catalyzed reactions play important roles in regulating various important life processes. In this section, the development of the transition state analog inhibitors for the isomerase-catalyzed reactions is illustrated with notable examples.

4.2.5.1 A Transition State Analog Inhibitor for the Pin1-catalyzed Isomerization Reaction

Pin1 is a peptidyl-prolyl isomerase and catalyzes the *cis–trans* isomerization of the phosphoSer/Thr-Pro peptide bond, as illustrated in Figure 4.8(A) with the phosphoSer-Pro substrate. The Pin1-catalyzed isomerization reaction plays a regulatory role in cell cycle progression and its inhibition would possess anti-cancer potential.

Figure 4.8(A) depicts the proposed chemical mechanism for the Pin1-catalyzed peptidyl-prolyl *cis–trans* isomerization.[10] The key mechanistic feature is the Pin1-catalyzed conformational distortion of the prolyl ring in the substrate to form a non-planar peptide bond in the transition state of the reaction. Based on this, the compound depicted in Figure 4.8(B) was designed as a potential transition state analog inhibitor and was indeed found to be a potent inhibitor for the Pin1-catalyzed isomerization reaction.[10] The design of this compound resulted from the replacement of the *cis–trans* isomerizable peptide bond in substrate with the indicated reduced peptide bond. The X-ray crystal structure of Pin1 with this compound bound at its active site revealed that the tertiary amine of the bound compound assumed the pyramidal conformation as depicted in the right-hand structure in Figure 4.8(B). This conformation would mimic the non-planar peptide bond conformation in the transition state depicted in Figure 4.8(A). Moreover, the observation suggests that Pin1 employs substrate conformational distortion as one strategy in its catalysis.

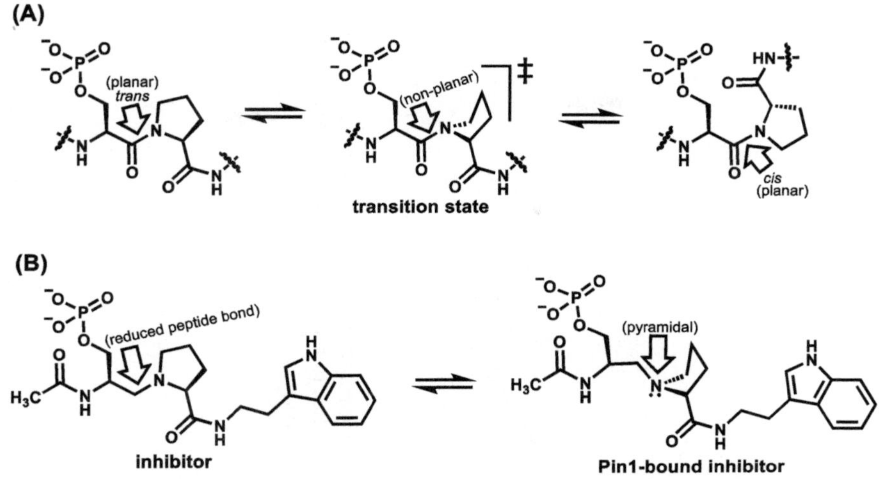

Figure 4.8 (A) The proposed chemical mechanism for the Pin1-catalyzed peptidyl-prolyl *cis–trans* isomerization reaction illustrated with the phosphoSer-Pro substrate. The non-planar distorted peptide bond in the proposed transition state and the planar peptide bond in both substrate and product are all indicated. (B) The chemical structure of a transition state analog inhibitor for the Pin1-catalyzed isomerization reaction, in which the *cis–trans* isomerizable peptide bond in the substrate of the reaction is changed to the indicated reduced peptide bond. The chemical structure on the right shows the tertiary amine pyramidal conformation which the inhibitor assumes when bound at the Pin1 active site and would mimic the non-planar distorted peptide bond conformation in the transition state depicted in (A).

4.2.5.2 A Transition State Analog Inhibitor for the Chorismate Mutase-catalyzed Isomerization Reaction

Chorismate mutase of the shikimate pathway catalyzes the conversion of chorismate to prephenate (Figure 4.9(A)), which plays an important role in the biosysnthesis of aromatic amino acids (*e.g.* tyrosine and phenylalanine) in non-mammalian species including bacteria, fungi, and higher plants. As an enzymatic reaction absent in mammals, the chorismate mutase-catalyzed reaction has been regarded as a potential therapeutic target for developing novel antibiotics, fungicides, and herbicides.

Figure 4.9(A) depicts the proposed chemical mechanism for the chorismate mutase-catalyzed reaction.[11] The transition state is presumed to possess the depicted bicyclic structure resulting from the enzyme-catalyzed conformational and configurational adjustment of the ring structure in substrate chorismate. Based on this mechanism, the compound depicted in Figure 4.9(B) was designed as a potential transition state analog inhibitor for the chorismate mutase-catalyzed reaction, with its bicyclic structure potentially mimicking the configuration of the transition state depicted in Figure 4.9(A).[11] This compound was found to be a potent inhibitor, lending support to the design rationale.

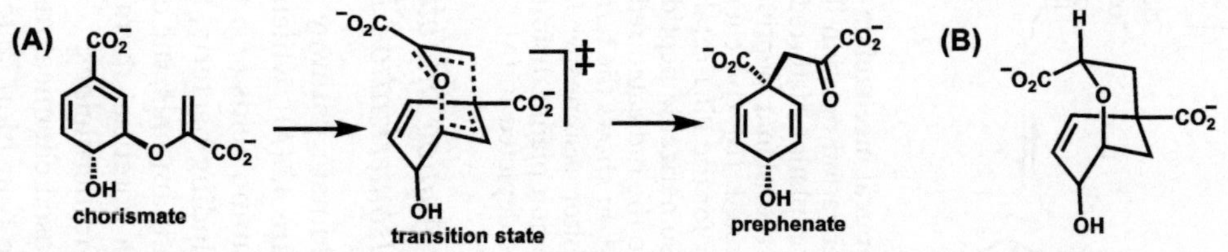

Figure 4.9 (A) The proposed chemical mechanism for the chorismate mutase-catalyzed reaction, with the proposed bicyclic transition state indicated. (B) The chemical structure of a transition state analog inhibitor for the chorismate mutase-catalyzed reaction, which would mimic the bicyclic configuration of the transition state depicted in (A).

4.2.6 Applications with Ligases

As stated above, ligases catalyze a variety of ligation reactions involved in important life processes. A few inhibitors (*e.g.* sulfonamide- and phosphonate-based) for the ligases MurD- and MurM-catalyzed reactions, which are involved in the biosynthesis of the peptidoglycan layer of the bacterial cell wall and are potential targets for the development of novel antibacterials, have been claimed in the literature to be transition state analog inhibitors. However, all these claimed inhibitory compounds were found to structurally mimic the catalytic tetrahedral intermediate instead of the preceding transition state, so they would be more appropriately classified as catalytic intermediate-based inhibitors. Therefore, transition state analog inhibitors have not yet been developed for ligase-catalyzed reactions.

References

1. A. E. Trunkfield, S. S. Gurcha, G. S. Besra and T. D. H. Bugg, *Bioorg. Med. Chem.*, 2010, **18**, 2651.
2. A. Simonian, A. Khomutov, T. Hyvonen, N. Grigorenko, T. Keinanen, J. Vepsalainen, L. Alhonen and J. Janne, *Nucleosides, Nucleotides Nucleic Acids*, 2007, **26**, 1245.
3. M. Welch, J. Chastang and M. Yarus, *Biochemistry*, 1995, **34**, 385.
4. K. Montin, C. Cervellati, F. Dallocchio and S. Hanau, *FEBS J.*, 2007, **274**, 6426.
5. T. M. Penning, M. E. Burczynski, J. M. Jez, H. K. Lin, H. Ma, M. Moore, K. Ratnam and N. Palackal, *Mol. Cell. Endocrinol.*, 2001, **171**, 137.
6. M. Degano, S. C. Almo, J. C. Sacchettini and V. L. Schramm, *Biochemistry*, 1998, **37**, 6277.
7. F. Wu, D. Grossenbacher and H. Gehring, *Mol. Cancer Ther.*, 2007, **6**, 1831.
8. F. Wu, P. Christen and H. Gehring, *FASEB J.*, 2011, **25**, 2109.
9. R. R. Somani, P. R. Rai and P. S. Kandpile, *Mini-Rev. Med. Chem.*, 2018, **18**, 1008.
10. G. G. Xu, Y. Zhang, A. Y. Mercedes-Camacho and F. A. Etzkorn, *Biochemistry*, 2011, **50**, 9545.
11. A. Mandal and D. Hilvert, *J. Am. Chem. Soc.*, 2003, **125**, 5598.

CHAPTER 5

(Photo)affinity Label and Covalent Inhibitor Design

5.1 Mode of Working

Following the initial reversible binding of a ligand (a substrate (or its analog) bearing an electrophilic group at an appropriate position) at the enzyme active site, if the electrophile on the ligand molecule is able to form a covalent bond with a nucleophile at the enzyme active site from an amino acid side chain or a coenzyme molecule, an irreversible structural/functional modification of the enzyme and its active site would then occur. Such a ligand is known as an affinity label for the enzyme because it exhibits an affinity to the enzyme active site *via* reversible binding before the subsequent irreversible binding (labeling) occurs at the same active site. One key consideration in the design of an affinity label is that, following its reversible binding at an enzyme active site, the electrophile on the ligand molecule would be within a covalent bonding distance from a nucleophile from the active site whose chemical reactivity would be compatible with that of the electrophile on the reversibly bound ligand.

Since the binding of an affinity label at an enzyme active site would entail a two-step process, *i.e.* the initial reversible binding followed by a subsequent irreversible covalent bonding, the off-target interaction of the ligand at bulk solution or with un-intended biomacromolecules would be likely insignificant despite the unfortunate occurrence of the reversible binding of the ligand with an un-intended target, due to the possible lack of the subsequent irreversible covalent bonding step which would be necessary for a significant off-target impact. However, in order to further minimize the off-target impact of an affinity label, its photo-controllable version (*i.e.* photo-affinity label) was developed by using a latent electrophilic group that can be photo-unmasked and subsequently binds irreversibly to the enzyme active site. Compared with an

Active Site-directed Enzyme Inhibitors: Design Concepts
By Weiping Zheng

Published by the Royal Society of Chemistry, www.rsc.org

affinity label, two additional enabling features of a successfully developed photo-affinity label would be: (1) being chemically inert at bulk solution and (2) a possibly diminished off-target ligand binding due to the presence on the molecule of electrosteric hindrance when binding with an off-target.

(Photo)affinity labels entertain a multitude of biochemical applications, such as the identification and structural characterization of an enzyme active site and the differentiation and classification of the homologous iso-enzymes within an enzyme family.

In view of the activity control of an enzyme-catalyzed reaction, the relevant affinity labels also behave as covalent inhibitors for the enzyme-catalyzed reaction and whose development has engendered a revitalized interest in transforming such reagents into potential therapeutic agents for human diseases. The modes of working for the (photo)affinity labeling of an enzyme active site and the covalent inhibition of the corresponding enzymatic reaction are illustrated in the following sections with respective notable examples.

5.2 Applications

As with the topic organization of Chapters 1–4, notable examples of (photo)affinity enzyme labels and covalent inhibitors for each of the six types of enzymes and the corresponding enzymatic reactions (*i.e.* transferases, oxidoreductases, hydrolases, lyases, isomerases, or ligases and the reactions they catalyze) are elaborated below.

5.2.1 Applications with Transferases

In this section, notable examples of the (photo)affinity enzyme labels and covalent inhibitors for a few transferase-catalyzed reactions are elaborated.

5.2.1.1 *An Affinity Label for Glutathione* S-*Transferase I*

Glutathione transferase I (GST I or ZmGSTF1), the most abundant GST in maize, is a member of the super-family of GSTs and catalyzes the covalent conjugation of glutathione (GSH) with various electrophilic substrate molecules *via* a thioether linkage. The GST I-catalyzed reaction plays a role in the detoxification of various electrophilic herbicides.

Figure 5.1(A) depicts the GST I-catalyzed nucleophilic substitution (a type of covalent conjugation reaction) between the nucleophilic side chain thiol (SH) of GSH and the electrophilic moiety present in an electrophilic substrate, leading to the formation of the depicted thioether-containing chemically neutral product.[1] Based on this, if an electrophilic moiety could be appended onto this reaction product at an appropriate position so that it could undergo a condensation reaction with a nearby nucleophile presented by the active site after the reversible binding of the electrophilic analog at the GST I active site, there would be then a covalent modification of the GST I active site. Figure 5.1(B) depicts such

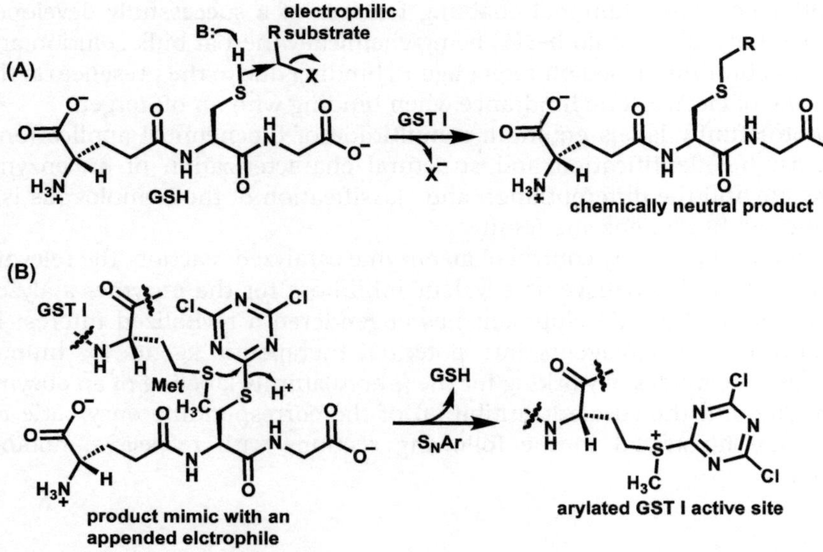

Figure 5.1 (A) The GST I-catalyzed nucleophilic substitution reaction between GSH and an electrophilic substrate. B:, a general base at the GST I active site. (B) The chemical structure of an electrophilic dichlorotriazinyl-containing product mimic serving as an affinity labeling reagent for GST I *via* the depicted S_NAr reaction.

an electrophilic analog and its mode of labeling of the GST I active site *via* an aromatic nucleophilic substitution (S_NAr) reaction between its electrophilic dichlorotriazinyl moiety and the nucleophilic side chain thioether of a methionine residue from the GST I active site. Moreover, this analog was found to be an efficient labeling reagent for not just GST I, but also other GSTs.[1]

5.2.1.2 *An Affinity Label for Lecithin Retinol Acyltransferase*

Lecithin retinol acyltransferase (LRAT) catalyzes the regioselective transesterification between a phosphatidylcholine derivative (lecithin) and a retinol analog (*e.g.* all-*trans*-retinol (vitamin A)), *i.e.* the acyl group transfer from the *sn*-1 position of lecithin to the primary hydroxyl of vitamin A, with the formation of lysolecithin and an all-*trans*-retinyl ester, as depicted in Figure 5.2(A).[2] The LRAT-catalyzed reaction plays a critical role in the biosynthesis of the visual pigment chromophore 11-*cis*-retinal in retinal pigment epithelium *via* the isomerohydrolase-catalyzed transformation of an all-*trans*-retinyl ester to 11-*cis*-retinol.

Since it was known that various thiol-directed compounds such as *p*-hydroxymercuribenzoate were able to irreversibly inhibit the LRAT-catalyzed reaction, the α-bromoacetate-based mimic of all-*trans*-retinyl ester depicted in Figure 5.2(B) was developed as a potential affinity labeling reagent for LRAT, since α-bromoacetate is also a known electrophilic thiol-directed functionality.[2]

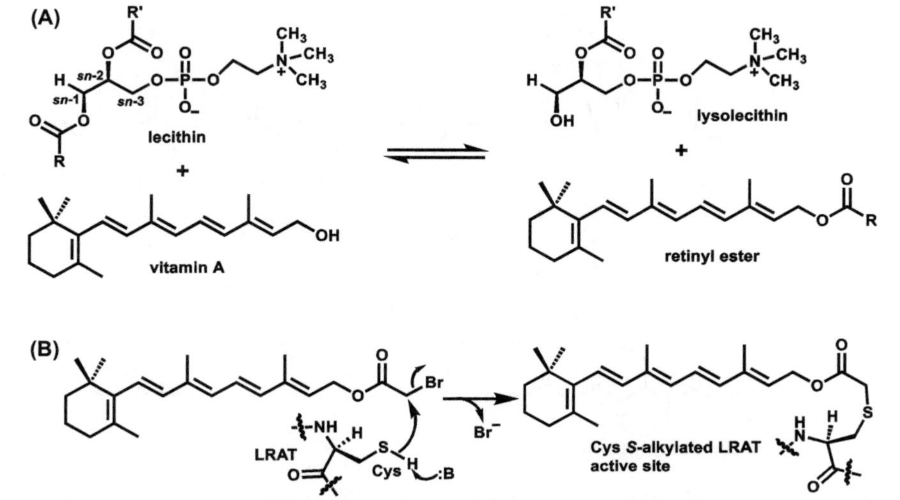

Figure 5.2 (A) The LRAT-catalyzed regioselective trans-esterification between lecithin and vitamin A (all-*trans*-retinol) with the formation of lysolecithin and retinyl ester. R and R′ are the fatty chains respectively at *sn*-1 and *sn*-2 positions. (B) The chemical structure of an electrophilic α-bromoacetate-based mimic of retinyl ester serving as an affinity labeling reagent for LRAT *via* the depicted cysteine side chain thiol alkylation. :B, a general base at the LRAT active site.

As shown and implied in Figure 5.2(B), the cysteine side chain thiol alkylation by the depicted α-bromoacetate-based compound would lead to an irreversible inhibition of the LRAT-catalyzed reaction *via* the depicted covalent labeling of LRAT. Indeed, the tritiated version of this compound was found to be able to label LRAT's catalytic subunit (molecular weight (MW) ∼ 25 kDa), as judged by sodium dodecyl sulfate-polyacrylamide gel electrophoresis.

5.2.1.3 *A Photo-affinity Label for Glutathione* S-*Transferase 4-4*

Similar to glutathione transferase I described in Section 5.2.1.1, glutathione *S*-transferase 4-4 (GST 4-4) is also able to catalyze the covalent conjugation of GSH with various electrophilic substrate molecules. The GST 4-4-catalyzed reaction also plays a role in detoxification of various electrophilic compounds including endogenous metabolites and exogenous xenobiotics.

Figure 5.3(A) depicts the GST 4-4-catalyzed nucleophilic substitution (a type of covalent conjugation reaction) between the nucleophilic side chain SH of GSH and the electrophilic moiety present in an electrophilic substrate, as described in Figure 5.1(A) for the GST I-catalyzed reaction, leading to the formation of the depicted thioether-containing chemically neutral product.[3] Based on this and the reasoning described in Section 5.2.1.1 on the design of the affinity label for GST I, the benzophenone-based compound depicted in Figure 5.3(B) was designed as a potential photo-affinity label for GST 4-4.[3] Benzophenone is a well known functionality whose ketone moiety can be

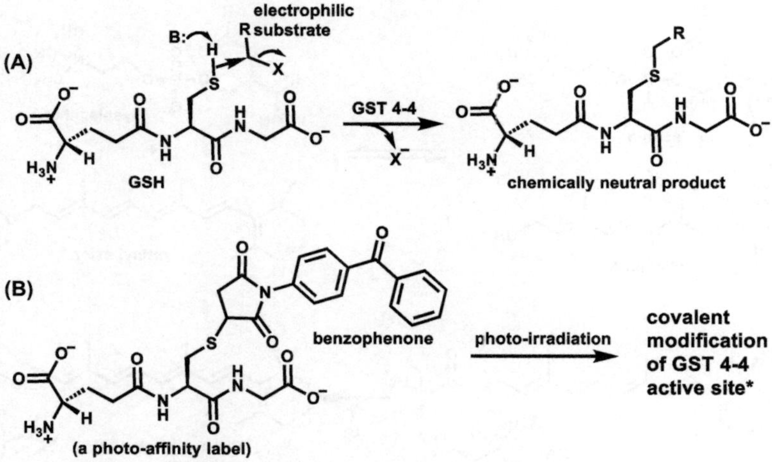

Figure 5.3 (A) The GST 4-4-catalyzed nucleophilic substitution reaction between GSH and an electrophilic substrate. B:, a general base at the GST 4-4 active site. This scheme is identical to that shown in Figure 5.1(A). (B) The chemical structure of a product mimic that contains a photo-activatable latent electrophile (*i.e.* benzophenone), which would serve as a photo-affinity labeling reagent for GST 4-4. The asterisk by "GST 4-4 active site" denotes that its covalent modification following photo-irradiation could be on an active site methionine residue by the benzophenone-derived highly electrophilic free radical (*i.e.* carbene).

converted to a free radical (*i.e.* carbene) upon photo-irradiation, and because of the high electrophilicity of a free radical that can interact essentially with anything nearby, even the inert C–H bond, the positioning of the benzophenone on a photo-affinity label could be more flexible than that for the electrophilic moiety on an affinity label. As compared to the interaction mode of an affinity label to an enzyme active site, that for the photo-affinity label would entail the addition of an intervening step between the initial reversible binding and the last step of irreversible covalent bonding, *i.e.* the photo-irradiation and thus photo-activation of the latent electrophilic functionality in a photo-affinity label. Of note, a GST 4-4 active site methionine residue was proposed to be the most probable site of modification by the benzophenone-derived highly electrophilic free radical. When the incubation of GST 4-4 and the designed photo-affinity labeling reagent was photo-irradiated with long-wavelength ultraviolet light, a time-dependent irreversible photo-inhibition of the GST 4-4-catalyzed reaction was observed, lending support to the design rationale.

5.2.1.4 *A Photo-affinity Label for Farnesyl Protein Transferase*

Farnesyl protein transferase (FPTase) catalyzes the farnesyl pyrophosphate (FPP)-dependent *S*-farnesylation (a type of prenylation) on specific cysteine side chains on several proteins such as the Ras proto-oncoprotein, as shown in Figure 5.4(A), with the driving force for this enzymatic reaction being the

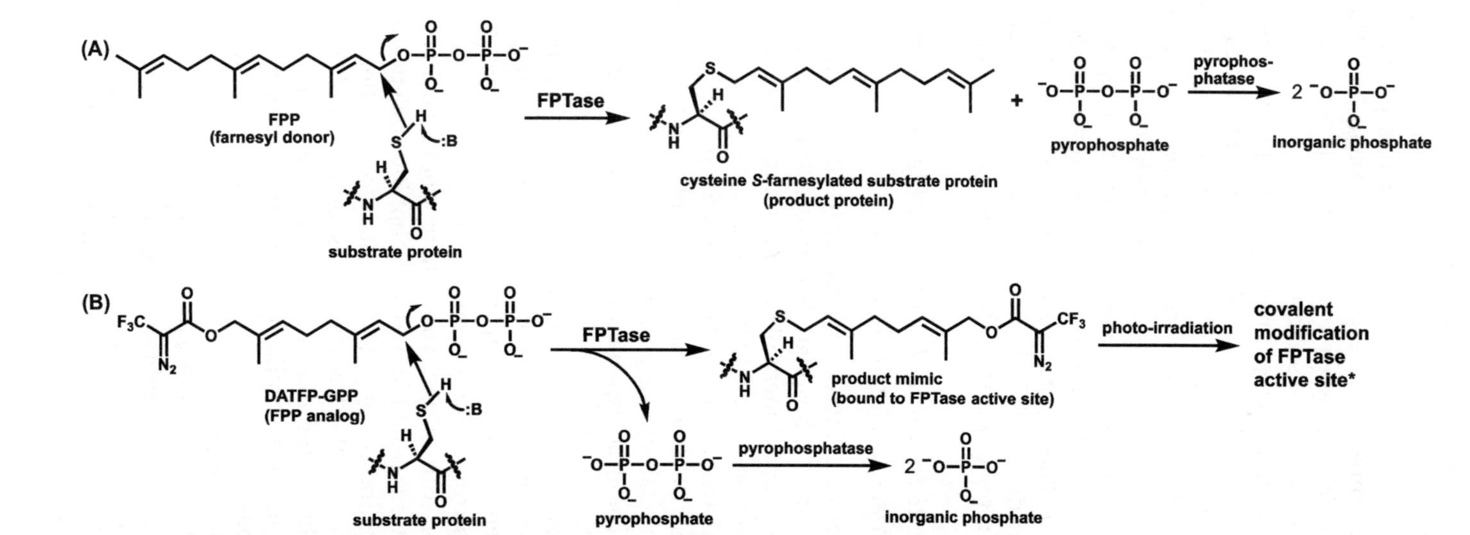

Figure 5.4 (A) The FPTase-catalyzed FPP-dependent cysteine side chain *S*-farnesylation on proteins. :B, a general base at the FPTase active site. (B) The FPTase-catalyzed processing of DATFP-GPP, a structural analog of FPP, with the formation of the depicted product which would covalently label FPTase active site upon photo-irradiation. :B, a general base at the FPTase active site. Note: the asterisk by "FPTase active site" denotes that its covalent modification following photo-irradiation was accomplished by the 2-diazo-3,3,3-trifluoropropionyloxy-derived highly electrophilic free radical (carbene).

pyrophosphatase-catalyzed hydrolysis of pyrophosphate product to inorganic phosphate.[4] The FPTase-catalyzed reaction plays an important regulatory role in cellular events proximal to cellular membranes (*e.g.* plasma membrane), with one notable case involving the proto-oncoprotein Ras. Therefore, the inhibitors for the FPTase-catalyzed reaction hold promise as therapeutic agents for human diseases such as cancer.

When 2-diazo-3,3,3-trifluoropropionyloxy-geranyl pyrophosphate (DATFP-GPP), a structural analog of FPP, was subjected to a FPTase catalytic activity assay, it was found to also serve as a co-substrate for the FPTase-catalyzed reaction, as depicted in Figure 5.4(B);[4,5] however, the resulting active site bound *S*-alkylated product seems to be able to covalently label the FPP-binding β-subunit of FPTase following photo-irradiation, as suggested from the photolysis experiment at 254 nm with a radio-labeled version of DATFP-GPP, *i.e.* [^{32}P]-DATFP-GPP. Following photo-irradiation, DATFP-GPP was also found to exhibit a time-dependent inhibition of the FPTase catalytic activity, which could be prevented from occurring by the presence of the co-substrate FPP.

5.2.1.5 A Covalent Inhibitor for the O-Acetylglucosamine Transferase-catalyzed Reaction

The intracellular enzyme *O*-acetylglucosamine (*O*-GlcNAc) transferase (OGT) can catalyze the uridine diphosphate (UDP)-GlcNAc-dependent *O*-GlcNAcylation (*i.e.* the covalent attachment of GlcNAc) on the side chain hydroxyl of specific serine or threonine residues on substrate proteins with the formation of UDP and the *O*-GlcNAcylated protein product, as depicted in Figure 5.5(A).[6,7] The OGT-catalyzed reaction plays an important role in regulating various crucial cellular processes, therefore, the inhibition of this enzymatic reaction could be exploited in developing therapeutic agents for human diseases such as cancer.

In view of inhibitor development, if we can design an electrophilic mimic of the reaction substrate (*i.e.* UDP-GlcNAc) so that, following its binding at OGT active site, there would present a nucleophile at the active site that would be spatially proximal and chemically compatible to the electrophilic moiety equipped on the mimic, then there would be likely a covalent interaction between electrophilic and nucleophilic moieties and the consequent covalent inhibition of the OGT-catalyzed reaction. Figure 5.5(B) depicts such a mimic of UDP-GlcNAc that was found in an *in vitro* assay to be an irreversible inhibitor for the OGT-catalyzed reaction *via* its covalent interaction through its electrophilic allylic chloride side arm with a cysteine residue side chain SH at the OGT active site.[6] However, due to the highly charged nature of this mimic, *i.e.* the dense negative charge on the pyrophosphate part of the molecule, it would be expected to be unable to cross the lipid bilayer of cellular membranes (*e.g.* plasma membrane), thus limiting its efficacy on the intracellular enzyme OGT. Therefore, to potentially circumvent this problem, the tetra-acetylated version of the GlcNAc mimic shown in Figure 5.5(B) was developed, hoping this compound would be cell

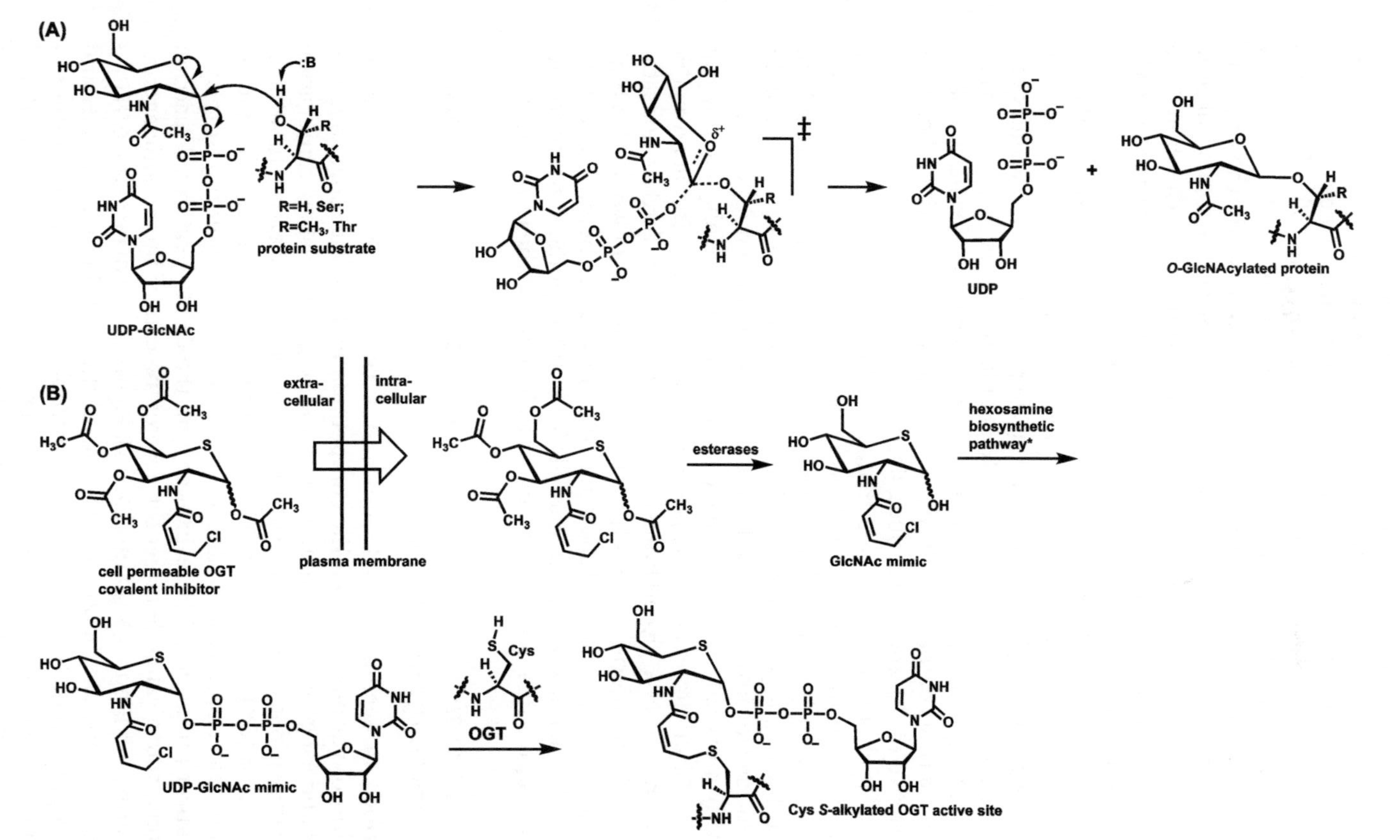

Figure 5.5 (A) The OGT-catalyzed *O*-GlcNAcylation of substrate proteins on specific serine or threonine side chains. :B, a general base at the OGT active site. (B) The processing of the indicated cell permeable tetra-acetylated version of the GlcNAc mimic, with the formation of the indicated UDP-GlcNAc mimic which would then be able to bind to the OGT active site and alkylate an active site cysteine residue side chain SH, leading to an irreversible inhibition of the OGT-catalyzed reaction. Note: the asterisk by "hexosamine biosynthetic pathway" denotes that this pathway represents just one possible pathway for the indicated transformation from the GlcNAc mimic to the UDP-GlcNAc mimic.

permeable and once inside cells, it would be converted to the desired UDP-GlcNAc mimic. It was indeed found to be cell permeable and was able to be transformed into the depicted UDP-GlcNAc mimic, which has been demonstrated in an *in vitro* assay to be an irreversible OGT inhibitor.[6] Specifically, once inside cells, the tetra-acetylated precursor would be hydrolyzed by intracellular esterases and the resulting GlcNAc mimic would then be converted to the depicted UDP-GlcNAc mimic by an intracellular biosynthetic pathway such as the hexosamine biosynthetic pathway.[6,7] Therefore, the tetra-acetylated compound would be a *bona fide* prodrug for the true OGT inhibitory compound, *i.e.* the UDP-GlcNAc mimic.

It should be noted that, according to the proposed occurrence of the oxonium cationic transition state depicted in Figure 5.5(A), the replacement of sulfur for the native oxygen in the pyranose ring of UDP-GlcNAc would deter the glycosyl transfer of the UDP hydrolysis side reactions from occurring, which could be due to the adoption of a non-productive conformation of the thio-pyranose ring compared with that of the native pyranose ring, therefore, the occurrence of the above-described desired covalent interaction would be maximized. It should be also noted that, even though UDP-GlcNAc is a universal sugar donor for many glycosylation reactions, the UDP-GlcNAc mimic was found to be fairly selective for the OGT-catalyzed reaction, which could be accounted for by the judicious selection and placement of the electrophilic allylic chloride side arm in the molecule, so that only those nucleophiles that are spatially proximal and chemically compatible would be alkylated by this covalent OGT inhibitor despite the possible occurrence of its initial reversible binding with multiple glycosyltransferase active sites. Of further note, the selection of allylic chloride as the electrophile in the inhibitor would also limit its side-reaction in the bulk solution due to its moderate chemical reactivity, however, it is chemically reactive enough for a nucleophilic substitution reaction with a proximal cysteine residue side chain SH at the OGT active site.

5.2.1.6 A Covalent Inhibitor for the Glutathione S-Transferase π-Catalyzed Reaction

As described in Sections 5.2.1.1 and 5.2.1.3, GSTs play a role in the detoxification of various electrophilic compounds including endogenous metabolites and exogenous xenobiotics *via* catalyzing the covalent conjugation of GSH with various electrophilic substrate molecules. In view of glutathione *S*-transferase π ($GSTP_{1-1}$), due to its over-expression in cancer cells and its catalysis on electrophilic anti-cancer agents, it would therefore also play a role in cancer promotion and resistance, and the $GSTP_{1-1}$-catalyzed reaction would constitute a valuable target for developing cancer therapeutic agents.

As described in Figures 5.1(A) and 5.3(A), Figure 5.6(A) depicts the $GSTP_{1-1}$-catalyzed condensation between the nucleophilic side chain SH of GSH and the electrophilic moiety present in an electrophilic substrate, with the formation of the depicted thioether-containing chemically neutral product.[8] Based on this and the reasoning described in Section 5.2.1.1 on the design of

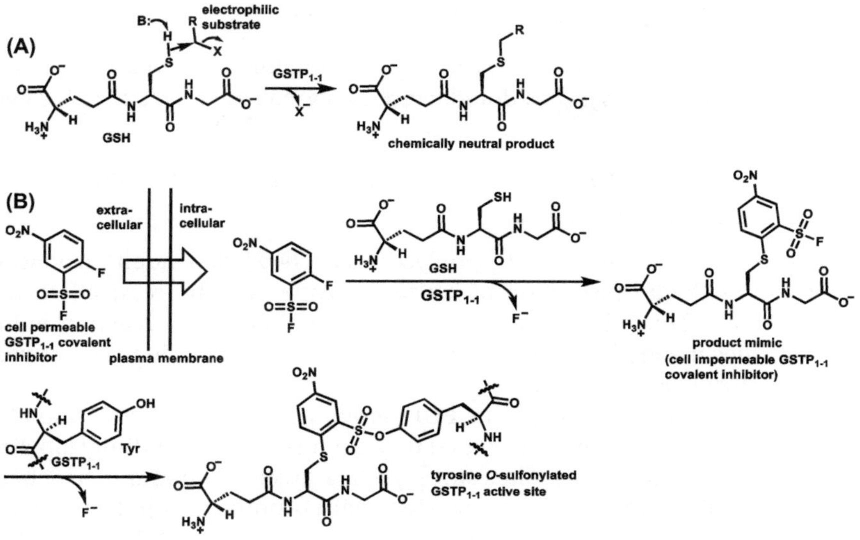

Figure 5.6 (A) The $GSTP_{1-1}$-catalyzed conjugation between GSH and an electrophilic substrate. B:, a general base at the $GSTP_{1-1}$ active site. This scheme depicts an example type of covalent conjugation reaction (*i.e.* nucleophilic substitution) and is identical to that shown in Figure 5.1(A). (B) The $GSTP_{1-1}$-catalyzed processing of the "cell permeable $GSTP_{1-1}$ covalent inhibitor" 2-fluoro-5-nitrobenzenesulfonyl fluoride following its cell permeation, with the *in situ* generation of the $GSTP_{1-1}$ covalent inhibitor "product mimic".

an affinity label for GST I, a similarly designed electrophilic mimic of the product of the $GSTP_{1-1}$-catalyzed reaction may also behave as a covalent inhibitor for the $GSTP_{1-1}$-catalyzed reaction. Figure 5.6(B) depicts such a compound in which sulfonyl fluoride (SO_2F) was used as the electrophile.[8] Of note, sulfonyl fluoride is a chemically moderately reactive electrophile, so its side-reaction in bulk solution would be limited, yet it is chemically reactive enough for a reaction with a proximal tyrosine side chain phenolic group at the $GSTP_{1-1}$ active site with the formation of the tyrosine *O*-sulfonylated $GSTP_{1-1}$.

As indicated in Figure 5.6(B), the $GSTP_{1-1}$ covalent inhibitor "product mimic" is tripeptidic in nature and a polar compound with a predictable low cell permeability. To circumvent this problem, a strategy based on the *in situ* generation of this cell impermeable inhibitory compound was pursued and realized, in that the intracellular 2-fluoro-5-nitrobenzenesulfonyl fluoride (*i.e.* the "cell permeable $GSTP_{1-1}$ covalent inhibitor" depicted in Figure 5.6(B)) was found to be converted to the inhibitor "product mimic" *via* the $GSTP_{1-1}$-catalyzed aromatic nucleophilic substitution reaction between 2-fluoro-5-nitrobenzenesulfonyl fluoride and the SH group of GSH.[8] Therefore, 2-fluoro-5-nitrobenzenesulfonyl fluoride is also a *bona fide* prodrug.

5.2.2 Applications with Oxidoreductases

In this section, notable examples of the (photo)affinity enzyme labels and covalent inhibitors for a few oxidoreductase-catalyzed reactions are elaborated.

5.2.2.1 *An Affinity Label for the NAD$^+$-dependent Formate Dehydrogenase*

Formate dehydrogenase (FDH) catalyzes the NAD$^+$-dependent oxidation of formate with the formation of carbon dioxide and the reduced NAD$^+$, *i.e.* NADH, as depicted in Figure 5.7(A).[9] The FDH-catalyzed reaction is important, since it constitutes the final step in the methanol oxidation pathway in the methylotrophic yeasts.

In an effort to help elucidate the FDH active site, the NAD$^+$/NADH analog 2′,3′-diketo-ADP (oADP) was evaluated to see if it could be an affinity labeling reagent for the FDH active site, since such a diketo compound had been known to be able to react with lysine side chain ε-amino group. By using the FDH from *Candida boidinii* as a model enzyme, it was found that oADP was capable of irreversibly inhibiting the FDH-catalyzed reaction *via* covalent modification of a catalytically important lysine (Lys) residue side chain ε-amino group at the NAD$^+$/NADH binding site in the form of a protonated Schiff base, as depicted in Figure 5.7(B).[9] Therefore, oADP could be an affinity label for the FDH active site, particularly the NAD$^+$/NADH binding site, since it is a NAD$^+$/NADH analog.

5.2.2.2 *An Affinity Label for Monoamine Oxidase-B*

Monoamine oxidase-B (MAO-B) catalyzes the flavin adenine dinucleotide (FAD)-dependent oxidative deamination of biogenic monoamines such as 5-hydroxytryptamine and dopamine (a catecholamine), as depicted in Figure 5.8(A) with phenylethylamine (an artificial substrate that MAO-B prefers). The MAO-B-catalyzed reaction plays an important role in regulating various cellular processes related to mood control. The inhibitors of the MAO-B-catalyzed reaction hold a therapeutic potential for modulating human mood disorders.

As shown in Figure 5.8(A), MAO-B is able to catalyze the deamination of phenylethylamine to form the depicted aldehydic product *via* the iminium cationic intermediate, with the concomitant reduction of FAD to FADH$_2$.[10] It is interesting to note that coenzyme FAD is covalently anchored at the MAO-B active site *via* the depicted cysteine (Cys) side chain SH at the 8α-carbon of FAD; moreover, there is a histidine (His) residue within this FAD binding region that is close to the afore-mentioned Cys residue, and whose side chain nucleophilic nitrogen could be the covalent attaching site for the aldehydic product of the MAO-B-catalyzed deamination of its selective inhibitor lazabemide.

Figure 5.8(B) depicts the MAO-B-catalyzed processing of lazabemide with the generation of the corresponding aldehydic intermediate which can then covalently modify MAO-B, likely *via* the depicted Schiff base formation with

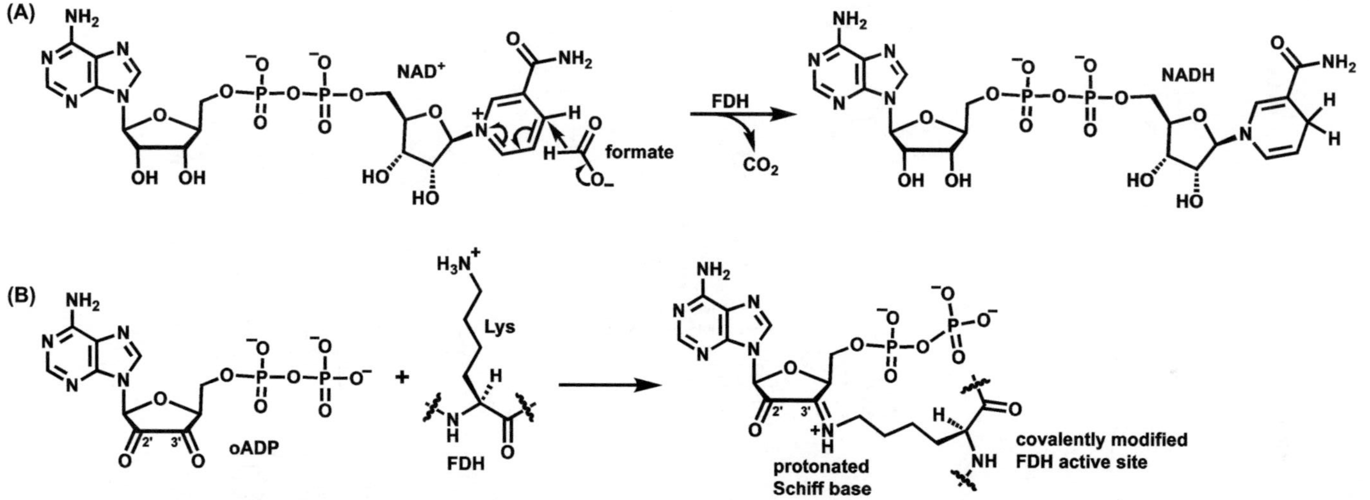

Figure 5.7 (A) The FDH-catalyzed NAD$^+$-dependent oxidation of formate to generate carbon dioxide and NADH. (B) The formation of the indicated protonated Schiff base between oADP and a lysine residue (Lys) at the NAD$^+$/NADH binding site of FDH, leading to its covalent labeling. oADP, 2′,3′-diketo-ADP. Note: only the 3′-modificaton by oADP is shown here, even though modification at the 2′ position would also be possible.

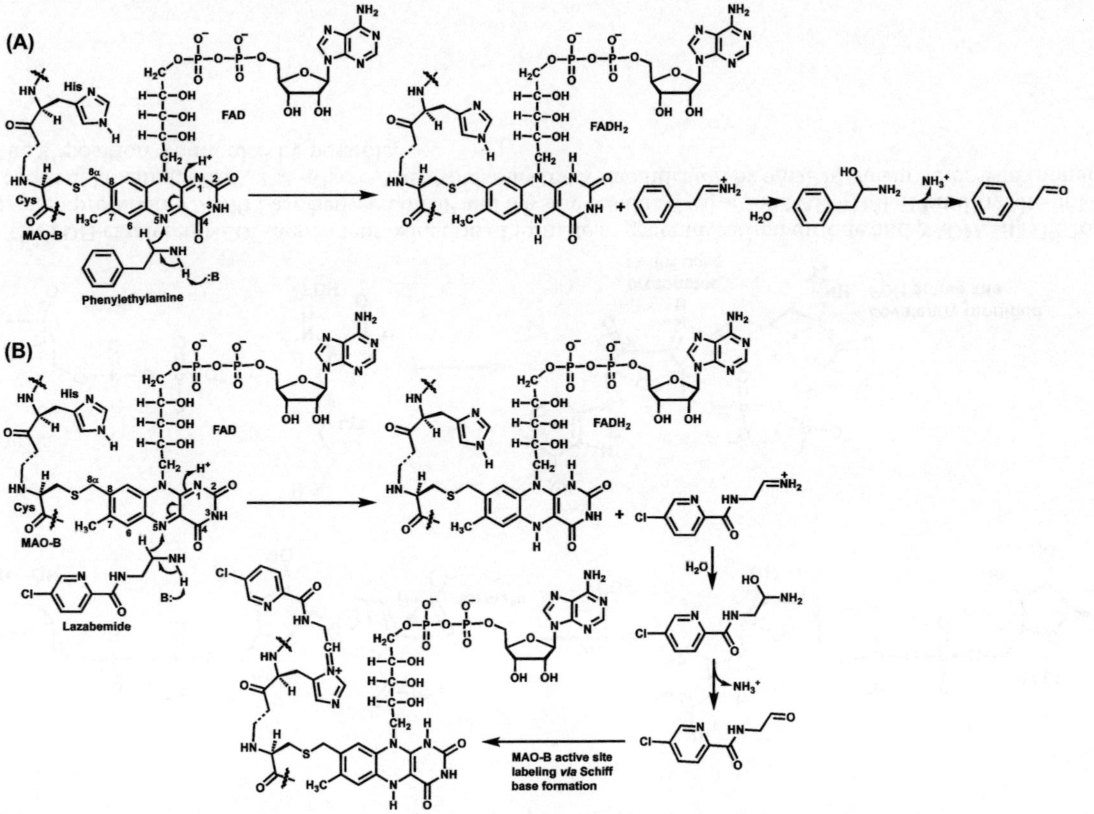

Figure 5.8 (A) The MAO-B-catalyzed FAD-dependent oxidative deamination of phenylethylamine to generate the depicted aldehydic product *via* the iminium cationic intermediate, with the concomitant reduction of FAD to $FADH_2$. :B, a general base at the MAO-B active site. (B) The MAO-B-catalyzed processing of lazabemide with the generation of the corresponding aldehydic intermediate which can then covalently modify MAO-B active site likely *via* the depicted Schiff base formation with the indicated His residue within the covalent FAD binding region at the MAO-B active site.

the afore-mentioned His residue within the covalent FAD binding region at the MAO-B active site, thus explaining the observed irreversible inhibition of the MAO-B-catalyzed reaction by lazabemide.[10] However, no MAO-B active site modification was observed with the aldehydic product formed from the substrate phenylethylamine. One possible explanation for this would be that the binding of the aldehydic product from phenylethylamine, which is structurally different from that formed from lazabemide, would engender a non-reactive orientation for this aldehydic product at the MAO-B active site, preventing it from forming a covalent adduct with MAO-B.

5.2.2.3 An Affinity Label for Aldose Reductase

Aldose reductase (AR) catalyzes the NADPH-dependent reduction of glucose to sorbitol, with the concomitant oxidation of NADPH to $NADP^+$, as shown in Figure 5.9(A).[11] The excessive production of sorbitol *via* this enzymatic reaction can lead to the development of various diabetic complications such as retinopathy, neuropathy, and nephropathy. Therefore, inhibitors for the AR-catalyzed reaction could be developed into a therapy for such pathological conditions.

Based on the AR inhibitor alrestatin, whose chemical structure is shown in Figure 5.9(B), its iodoacetamido analog depicted in Figure 5.9(C) has been developed and whose use as an affinity label has been instrumental to the identification of an alternative binding site on AR for various structurally diverse AR inhibitors.[11]

5.2.2.4 A Photo-affinity Label for the $NADP^+$-dependent Isocitrate Dehydrogenase

As described in Section 3.2.2.2 and Figure 3.7(A), isocitrate dehydrogenase (IDH) catalyzes the transformation of isocitrate to α-ketoglutarate, as also shown in Figure 5.10(A) together with the proposed chemical mechanism for the IDH-catalyzed reaction.[12] As the third reaction of the citric acid cycle, the IDH-catalyzed reaction plays an important role in regulating cellular metabolism, and its inhibitors could potentially be valuable research tools and therapeutic agents for human metabolic disorders. Figure 5.10(B) depicts the chemical structure of a compound designed as a benzophenone-based photo-affinity label for the coenzyme binding site on IDH, since this compound is structurally similar to coenzymes $NADP^+$ and NADPH engaged in the IDH-catalyzed reaction.[12] Following photo-irradiation, the benzophenone carbonyl would produce a highly electrophilic species (*i.e.* carbene) that can be inserted into any covalent bond nearby.

5.2.2.5 A Photo-affinity Label for the Peroxisomal Acyl-CoA Oxidase

A peroxisome is an organelle that harbors fatty acid β-oxidation enzymes and plays a vital role in regulating metabolic pathways in which fatty acid β-oxidation is an essential part. As depicted in Figure 5.11(A), the

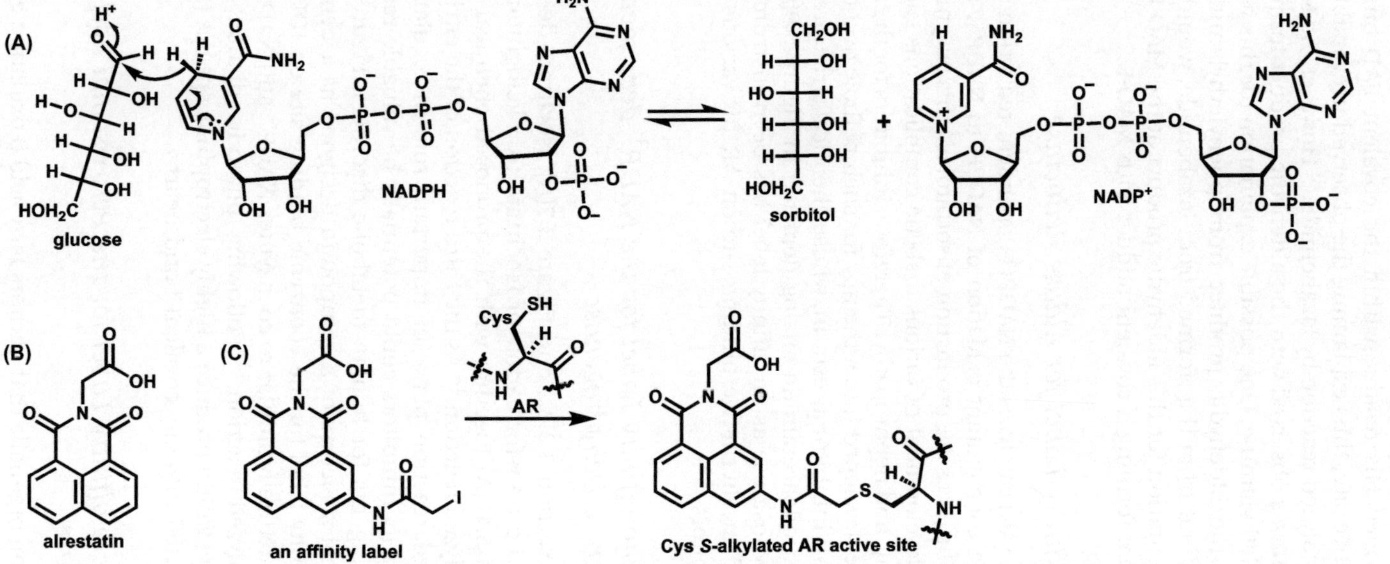

Figure 5.9 (A) The aldose reductase (AR)-catalyzed NADPH-dependent reduction of glucose to sorbitol, with the concomitant oxidation of NADPH to $NADP^+$. (B) The chemical structure of the AR inhibitor alrestatin. (C) The covalent labeling of AR by the depicted affinity label *via* covalent modification on an amino acid side chain, such as the depicted side chain *S*-alkylation on a cysteine (Cys) residue.

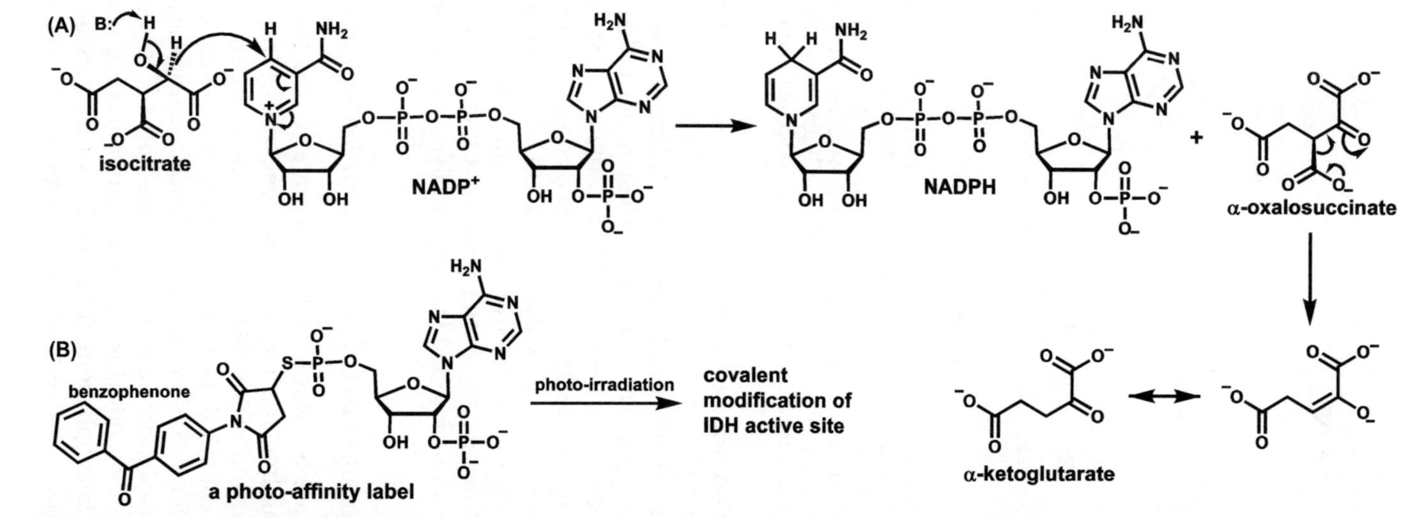

Figure 5.10 (A) The proposed chemical mechanism for the IDH-catalyzed reaction. B:, a general base at the IDH active site. This scheme is essentially same as that in Figure 3.7(A). (B) The chemical structure of the designed benzophenone-based photo-affinity label which structurally mimics the coenzymes $NADP^+$ and NADPH of the IDH-catalyzed reaction.

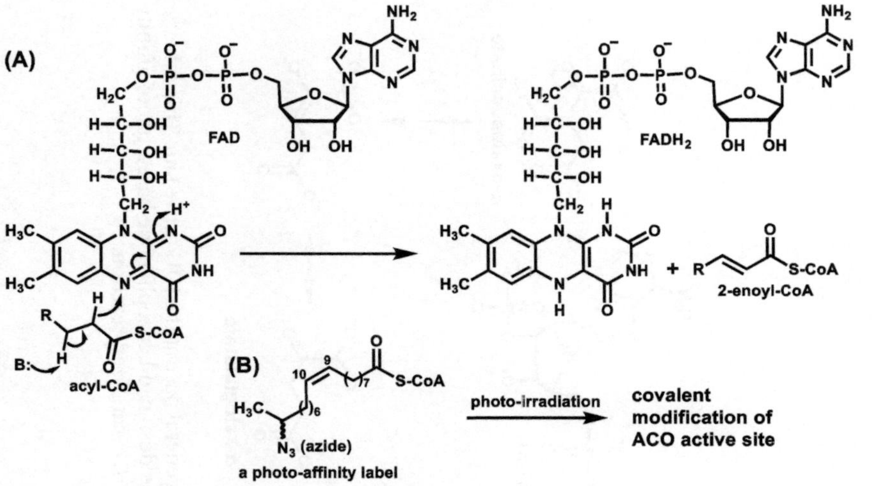

Figure 5.11 (A) The proposed chemical mechanism for the acyl-CoA oxidase (ACO)-catalyzed conversion of an acyl-CoA to the corresponding 2-enoyl-CoA, with the concomitant reduction of FAD to $FADH_2$. R, an aliphatic carbon chain; B:, a general base at the ACO active site. (B) The chemical structure of the acyl-CoA substrate mimic 12-azido-oleoyl-CoA that was designed as an azide-based photo-affinity label.

peroxisomal acyl-CoA oxidase (ACO) catalyzes the FAD-dependent conversion of an acyl-CoA to the oxidative product 2-enoyl-CoA with the concomitant reduction of FAD to $FADH_2$, which constitutes the first step of the peroxisomal fatty acid β-oxidation pathway.[13] In order to help define the active site of ACO, the compound depicted in Figure 5.11(B) (*i.e.* the acyl-CoA substrate mimic 12-azido-oleoyl-CoA) was designed as an azide-based photo-affinity label for ACO.[13] It was found to be recognized as an ACO substrate in darkness and to covalently modify the ACO active site following photo-irradiation, under which the azide functionality would produce a highly electrophilic species (*i.e.* nitrene) that can be inserted into any covalent bond nearby.

5.2.2.6 A Photo-affinity Label for Steroid 5α-Reductase Isozyme-1

Steroid 5α-reductase isozyme-1 (5αR-1) can catalyze the NADPH-dependent reduction of the androgenic ligand testosterone to dihydrotestosterone with concomitant oxidation of NADPH to $NADP^+$, as depicted in Figure 5.12(A) for the proposed chemical mechanism.[14] In order to help define the active site of 5αR-1, the structural mimic of dihydrotestosterone was designed as a benzophenone-based photo-affinity label for 5αR-1.[14] It was found to be capable of covalently modifying the active site following photo-irradiation, under which the benzophenone functionality would produce a highly electrophilic species (*i.e.* carbene) that can be inserted into any nearby covalent bond.

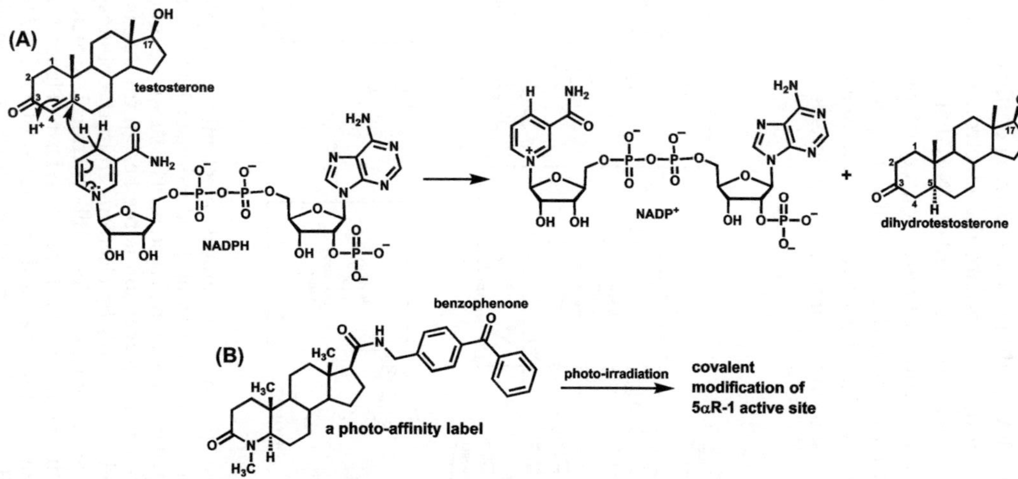

Figure 5.12 (A) The proposed chemical mechanism for the 5αR-1-catalyzed NADPH-dependent reduction of testosterone to dihydrotestosterone. (B) The chemical structure of a structural mimic of dihydrotestosterone designed as a photo-affinity label for 5αR-1.

5.2.2.7 A Covalent Inhibitor for the Ferredoxin-NADP$^+$ Reductase-catalyzed Reaction

Ferredoxin-NADP$^+$ reductase (FNR) catalyzes the oxidation of ferredoxin (the reduced Fe^{2+} form) with the concomitant reduction of NADP$^+$ to generate the reducing equivalent NADPH, as depicted in Figure 5.13(A).[15,16] Since the iron–sulfur center of ferredoxin is a one-electron carrier, the coenzyme flavin adenine dinucleotide (FAD) and its two reducing equivalents (*i.e.* FADH$^\bullet$ and FADH$_2$) are used in nature to mediate the electron transfer between ferredoxin and the two-electron carrier NADP$^+$/NADPH system, as depicted in Figure 5.13(A). Of note, the ultimate source of electrons for the FNR-catalyzed reaction in photosynthetic organisms is the photosystem I, as indicated in Figure 5.13(A). The FNR-catalyzed reaction is not present in mammals, yet is present and important in species such as the malaria

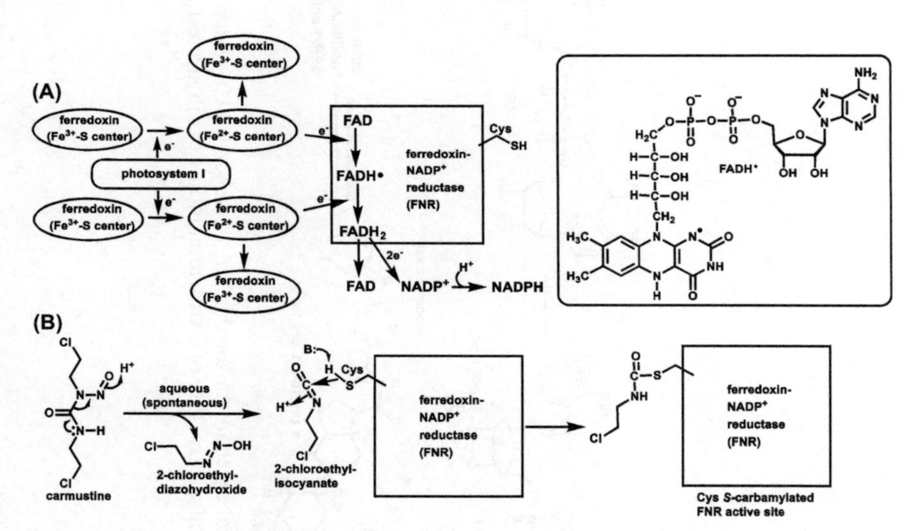

Figure 5.13 (A) The proposed chemical mechanism for the FNR-catalyzed oxidation of ferredoxin and reduction of NADP$^+$, intervened by the FAD/FADH$^\bullet$/FADH$_2$ redox system. The substrate protein ferredoxin for the FNR-catalyzed reaction is the Fe^{2+} form that is reduced from its oxidized (Fe^{3+}) form by electrons from the photosystem I in photosynthetic organisms. The full chemical structure of FADH$^\bullet$ is shown in the boxed inset; please refer to Figure 5.11 and 3.7 in this book for the full chemical structures of FAD, FADH$_2$, NADP$^+$, and NADPH. (B) The chemical structure and mode of irreversible FNR inhibition of the FDA-approved nitrosourea-based anti-cancer DNA-alkylator carmustine. As shown, carmustine gets degraded spontaneously in aqueous solution to 2-chloroethyl-diazohydroxide and 2-chloroethyl-isocyanate, with the latter compound being able to carbamylate a catalytic cysteine side chain SH in the FNR active site, leading to an irreversible inhibition of the FNR-catalyzed reaction. B:, a general base at the FNR active site.

parasite *Plasmodium falciparum*, therefore, inhibitors of the FNR-catalyzed reaction could be potentially developed into novel antimalarials.

Figure 5.13(B) depicts the chemical structure and mode of irreversible FNR inhibition of carmustine, which is an FDA-approved nitrosourea-based anti-cancer DNA-alkylating agent.[15] As shown, carmustine is chemically unstable in aqueous solution and gets degraded spontaneously to 2-chloroethyl-diazohydroxide and 2-chloroethyl-isocyanate. While the former compound mediates carmustine's cytostatic activity, the latter compound is able to carbamylate a catalytic cysteine side chain SH in the FNR active site, leading to an irreversible inhibition of the FNR-catalyzed reaction.

5.2.2.8 A Covalent Inhibitor for the Glyceraldehyde 3-Phosphate Dehydrogenase-catalyzed Reaction

Glyceraldehyde 3-phosphate dehydrogenase (GAPDH) is a key enzyme in the glycolysis pathway in all known living organisms, and catalyzes the $NADP^+$-dependent oxidative phosphorolysis of the pathway intermediate glyceraldehyde 3-phosphate (GAP), leading to the formation of the intermediate 1,3-bisphosphoglycerate (1,3-BPG) in the same pathway, as shown in Figure 5.14(A).[17,18] Given the importance of this pathway, especially for the survival of those cells that are devoid of the aerobic pathway for ATP generation, *e.g.* solid cancer cells undergoing the Warburg effect and the malaria parasite *Plasmodium falciparum*, the inhibitors for the glycolysis pathway such as those against the GAPDH-catalyzed reaction would hold a potential for the development of novel chemotherapeutic agents against cancer and parasitic infection.

Figure 5.14(B) depicts a 3-bromoisoxazoline-based covalent inhibitor for the GAPDH-catalyzed reaction.[17,18] As shown, this compound is able to interact with and alkylate a catalytic cysteine side chain SH in the GAPDH active site, thus leading to an irreversible inhibition of the GAPDH-catalyzed reaction.

5.2.2.9 A Covalent Inhibitor for the Aldehyde Oxidase-catalyzed Reaction

The mammalian cytosolic enzyme aldehyde oxidase (AO) catalyzes the oxidation of aldehydes and nitrogen-containing compounds, which plays an important role in the metabolism of xenobiotics and drugs. The active site of AO contains a square-pyramidal molybdenum (Mo) pyranopterin cofactor that in this particular case is equipped with three sulfur atoms, one oxygen atom, and one water-derived catalytically labile hydroxyl group, as indicated in Figure 5.15(A).[19–21] As shown with a nitrogen-containing compound, this enzymatic reaction proceeds *via* a nucleophilic attack of the free hydroxyl of the Mo cofactor onto the sp^2 carbon of the substrate, which is concerted with the transfer of the hydride from this sp^2 carbon onto the free sulfur of the Mo cofactor, giving rise to the enaminium product and the coordination-deficient

Figure 5.14 (A) The proposed chemical mechanism for the GAPDH-catalyzed NADP$^+$-dependent oxidative phosphorolysis of the glycolysis pathway intermediate GAP to form the intermediate 1,3-BPG in the same pathway. Cys, a catalytic cysteine at the GAPDH active site; :B, a general base at the GAPDH active site. (B) The chemical structure and mode of irreversible GAPDH inhibition of the indicated 3-bromoisoxazoline-based covalent inhibitor *via* alkylation of the catalytic cysteine side chain SH at the GAPDH active site, leading to an irreversible inhibition of the GAPDH-catalyzed reaction.

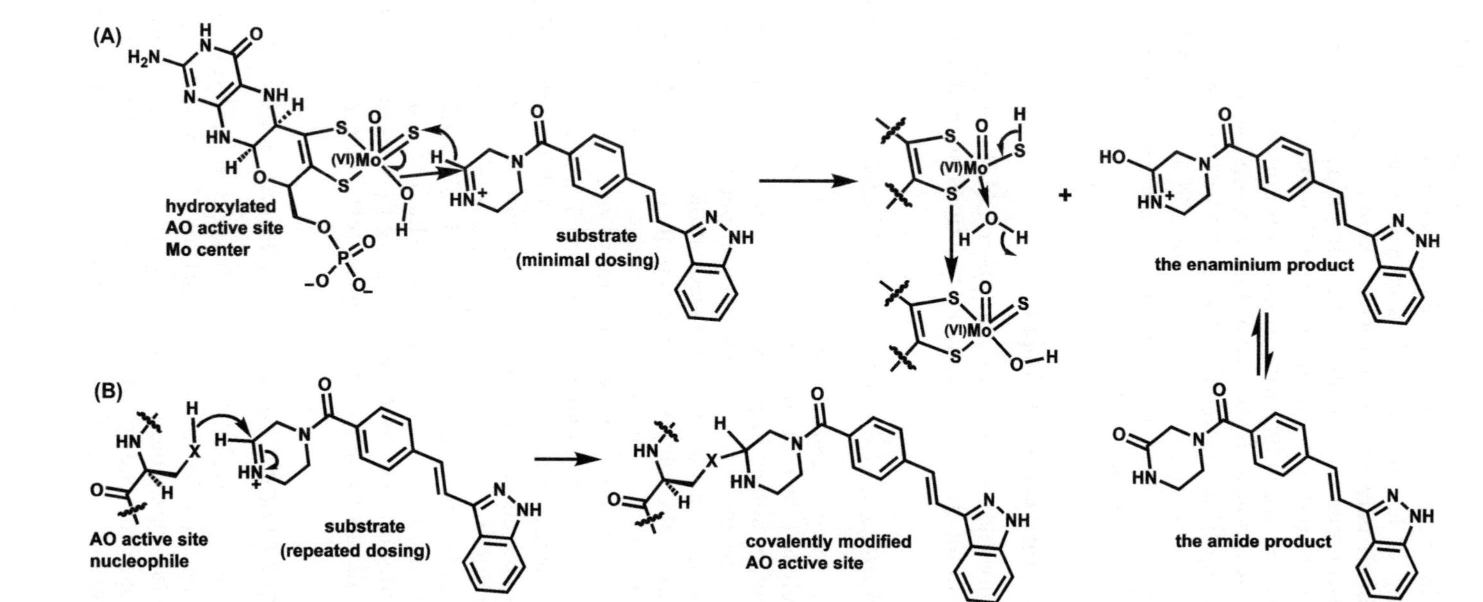

Figure 5.15 (A) The proposed concerted chemical mechanism for the AO-catalyzed oxidation of a nitrogen-containing compound under minimal dosing setting. Mo, molybdenum. (B) The proposed chemical mechanism accounting for the experimentally observed covalent irreversible inhibition of the AO-catalyzed reaction under repeated dosing setting. The nucleophile in this mechanism could be that from an AO active site catalytic residue.

Mo cofactor. The former is then tautomerized to the amide product and the latter is re-coordinated with water.[19] It should be noted that the chemical mechanism shown in Figure 5.15(A) is a concerted mechanism which has been supported by density functional theory calculations.

It should be also noted that the nitrogen-containing compound shown in Figure 5.15(A) was found to be a substrate under minimal dosing setting, however, interestingly, when its concentration was enhanced under repeated dosing setting it was found to be a covalent irreversible inhibitor for the AO-catalyzed reaction.[22] As depicted in Figure 5.15(B), the reaction between the sp^2 carbon of the compound and a nucleophile from the AO active site forming the depicted azaketal-like adduct could account for the observed covalent and irreversible inhibition.

5.2.3 Applications with Hydrolases

In this section, notable examples of the (photo)affinity enzyme labels and covalent inhibitors for hydrolase-catalyzed reactions are elaborated.

5.2.3.1 An Affinity Label for the New Delhi Metallo-β-lactamase-1

The New Delhi metallo-β-lactamase-1 (NDM-1) catalyzes the metal-dependent hydrolytic β-lactam ring opening of the β-lactam anti-bacterials such as chromacef, as shown in Figure 5.16(A), and plays an essential role in the emergence of resistance to the β-lactam anti-bacterials.[23] This bi-nuclear enzyme uses two Zn^{2+} ions to activate a water molecule and simultaneously further polarize the carbonyl group at the scissile position. As indicated, the depicted active site lysine residue is not involved in the chemistry of this enzymatic hydrolytic reaction, but has a putative role in substrate binding at the NDM-1 active site. Yet, as shown in Figure 5.16(B), this lysine residue plays a central role in mediating the covalent modification of the NDM-1 active site *via* its side chain's covalent modification by the depicted affinity label, which is structurally binary and composed of the covalently linked carbamate and carbonate moieties.[23] The availability of such an affinity label would facilitate the active site analysis, which would facilitate the development of inhibitors for the NDM-1-catalyzed β-lactam ring opening of β-lactam anti-bacterials to tackle the bacterial resistance problem.

5.2.3.2 An Affinity Label for S*-Adenosylhomocysteine Hydrolase*

As described in Section 1.2.3.5, *S*-adenosylhomocysteine hydrolase (AdoHcyase) catalyzes the thioether hydrolysis reaction of *S*-adenosylhomocysteine (AdoHcy) to form homocysteine and adenosine, as depicted in Figure 5.17(A),[24] and the AdoHcyase-catalyzed reaction plays an important role in regulating cellular (patho)physiological processes and represents a potential therapeutic target for human diseases.

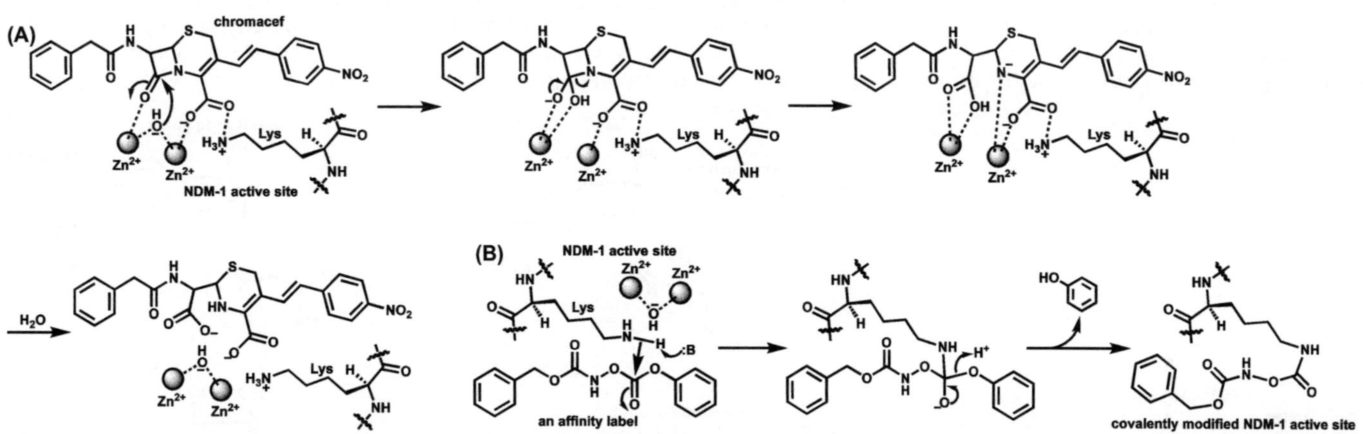

Figure 5.16 (A) The proposed chemical mechanism for the New Delhi metallo-β-lactamase-1 (NDM-1)-catalyzed hydrolysis of the β-lactam anti-bacterial chromacef. Lys, a lysine residue at NDM-1 active site. (B) The proposed chemical mechanism for the covalent modification of this lysine side chain at NDM-1 active site by a structurally binary affinity labeling reagent composed of covalently linked carbamate and carbonate moieties. As indicated, the catalytic bi-nuclear (bi-zinc) center is not involved in the catalytic processing of the affinity label. :B, a general base at the NDM-1 active site.

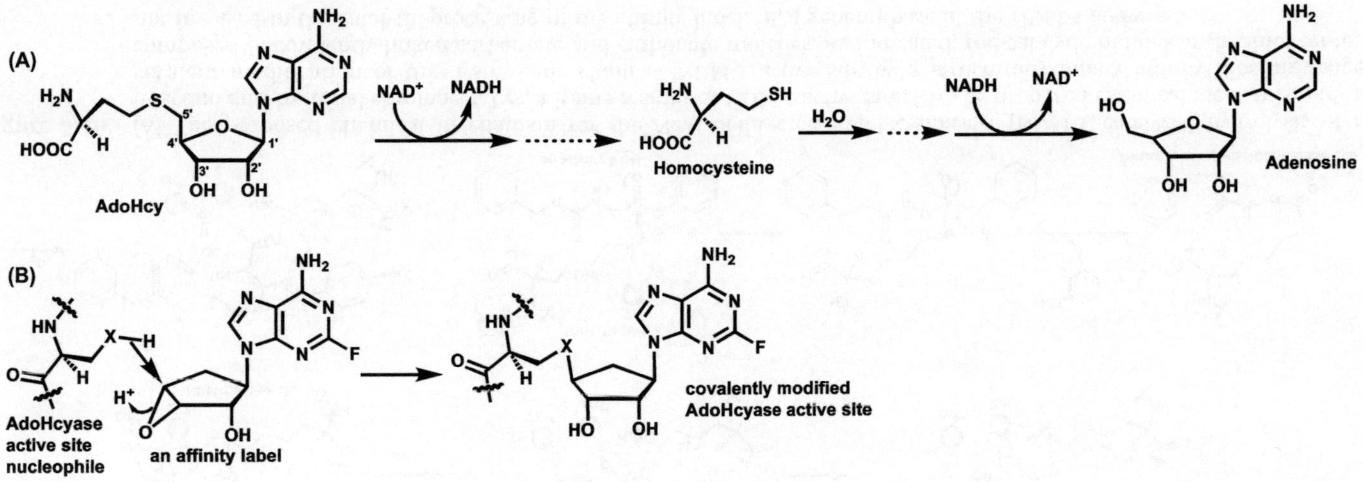

Figure 5.17 (A) A simplified version of the proposed chemical mechanism for the *S*-adenosylhomocysteine hydrolase (AdoHcyase)-catalyzed thioether hydrolysis reaction of *S*-adenosylhomocysteine (AdoHcy) to form homocysteine and adenosine. The fuller version of this mechanistic scheme is shown in Figure 1.20. Please see other figures (*e.g.* Figure 3.7) in this book for the full chemical structures of NAD^+ and NADH. (B) The proposed chemical mechanism accounting for the experimentally observed covalent irreversible inhibition of the AdoHcyase-catalyzed reaction by the indicated affinity label which is a structural analog of AdoHcy and adenosine. The nucleophile in this mechanism could be that from an AdoHcyase active site catalytic residue.

Figure 5.17(B) depicts a proposed chemical mechanism which could account for the experimentally observed covalent irreversible inhibition of the AdoHcyase-catalyzed reaction by the indicated affinity label *via* the covalent modification of a nucleophile from an AdoHcyase active site catalytic residue.[24] Of note, this affinity label is a structural analog of the substrate AdoHcy and the product adenosine for the AdoHcyase-catalyzed reaction.

5.2.3.3 A Photo-affinity Label for AdoHcyase

As described in Sections 1.2.3.5 and 5.2.3.2, AdoHcyase catalyzes the thioether hydrolysis reaction of AdoHcy to afford homocysteine and adenosine, as depicted in Figure 5.18(A).[25] And, as afore-mentioned, the AdoHcyase-catalyzed reaction plays an important role in regulating cellular (patho)physiological processes and represents a potential therapeutic target for human diseases.

Figure 5.18(B) depicts the chemical structure of 8-azido-adenosine, which is a close structural analog of the reaction product adenosine from the AdoHcyase-catalyzed reaction (see Figure 5.18(A)) and has been experimentally observed to be a photo-affinity labeling reagent for AdoHcyase.[25] Presumably, 8-azido-adenosine would bind to the AdoHcyase active site in darkness and covalently modify the AdoHcyase active site following photo-irradiation, under which the azide functionality would afford the highly electrophilic nitrene that can be inserted into any proximal covalent bond. The availability of such a labeling reagent would promote the analysis of the active site of AdoHcyase.

5.2.3.4 A Photo-affinity Label for Juvenile Hormone Epoxide Hydrolase

Juvenile hormone epoxide hydrolase (JHEH) catalyzes the conversion of juvenile hormone (JH) to its di-hydroxyl product, as shown in Figure 5.19(A).[26] The JHEH-catalyzed reaction plays a crucial role in regulating the titer of insect JH. It should be noted that this enzymatic reaction is still covered in this section despite its being a hydration reaction instead of a hydrolytic reaction. The chemical mechanism of the JHEH-catalyzed reaction is shown in Figure 5.19(A), in which the water molecule is depicted to nucleophilically attack the less sterically hindered carbon.[26]

Figure 5.19(B) depicts the chemical structure of a close structural analog of the JH molecule (*i.e.* the JHEH reaction substrate) depicted in Figure 5.19(A), which has been designed as an azide-based photo-affinity labeling reagent for JHEH.[26] Such an active site-directed reagent has been demonstrated to be useful in the identification and analysis of the JHEH active site. Following binding to the JHEH active site in darkness, the compound would presumably covalently modify the JHEH active site in response to photo-irradiation under which the azide functionality would afford the highly electrophilic nitrene that can be inserted into any proximal covalent bond.

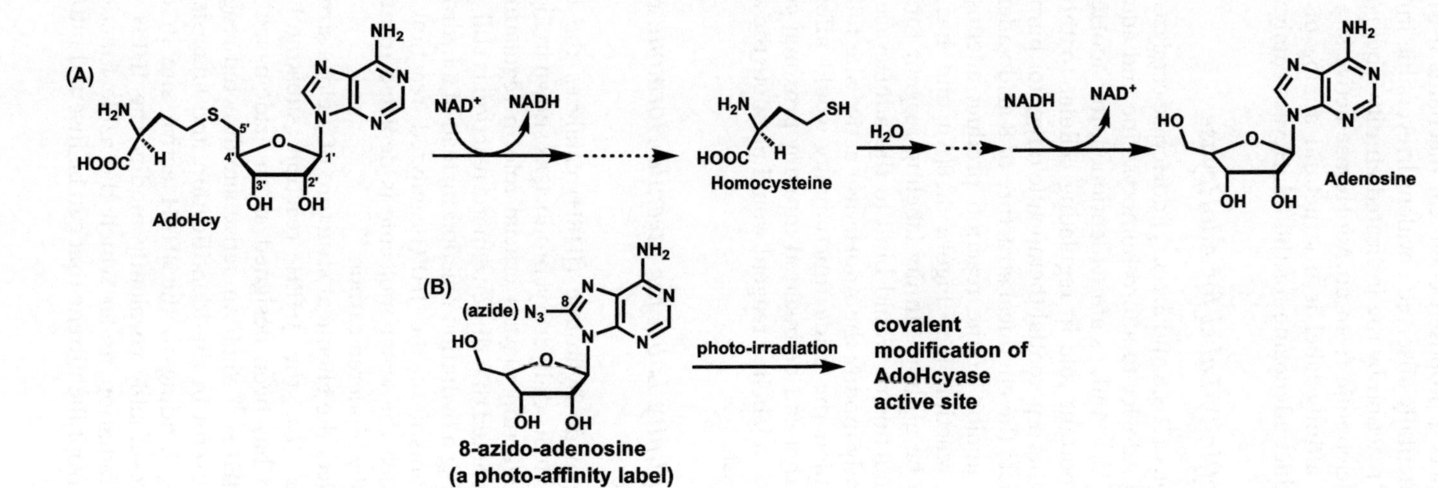

Figure 5.18 (A) The proposed chemical mechanism for the *S*-adenosylhomocysteine hydrolase (AdoHcyase)-catalyzed thioether hydrolysis reaction of *S*-adenosylhomocysteine (AdoHcy) to afford homocysteine and adenosine. This version of the mechanistic scheme is identical to that shown in Figure 5.17(A). (B) The chemical structure of a close structural analog of adenosine, *i.e.* 8-azido-adenosine, that was designed as an azide-based photo-affinity label for AdoHcyase.

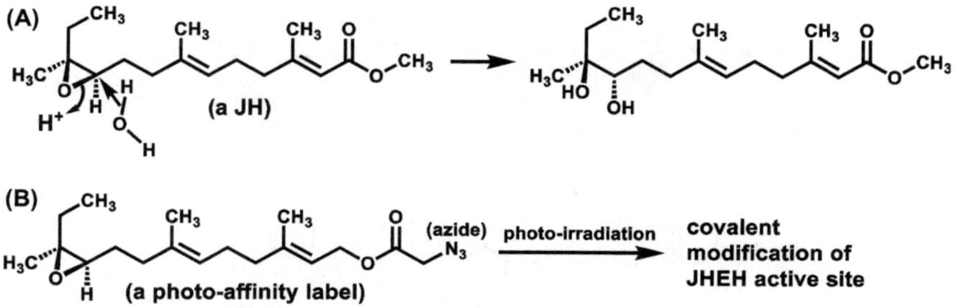

Figure 5.19 (A) The proposed chemical mechanism for the juvenile hormone epoxide hydrolase (JHEH)-catalyzed hydration reaction of a juvenile hormone (JH) molecule to its di-hydroxyl product. In this mechanistic scheme, a water molecule is depicted to nucleophilically attack the less sterically hindered epoxide carbon. (B) The chemical structure of a close structural analog of the JH substrate depicted in Figure 5.19(A), which was designed as an azide-based photo-affinity label for JHEH.

5.2.3.5 *A Covalent Inhibitor for the Severe Acute Respiratory Syndrome Coronavirus 2 Main Protease-catalyzed Reaction*

The severe acute respiratory syndrome coronavirus 2 (SARS-CoV-2) main protease (M^{pro}) is a cysteine protease that catalyzes the peptide bond hydrolysis of the viral poly-protein to afford several different mature proteins that are important for viral replication, therefore, the M^{pro}-catalyzed reaction plays an essential role in the life cycle of the SARS-CoV-2 virus and has been regarded as a therapeutic target for developing novel treatments to combat the SARS-CoV-2 pandemic.

Figure 5.20(A) depicts the proposed chemical mechanism for the M^{pro}-catalyzed reaction that proceeds through the two depicted covalent tetrahedral catalytic intermediates.[27] Therefore, if the catalytic cysteine side chain SH could be covalently and stably modified this enzymatic reaction could be irreversibly inhibited. Figure 5.20(B) depicts such an inhibitory compound that bears an electrophilic α-dichloroacetamide warhead upon a structural mimic of the M^{pro} reaction substrate.[27] As shown, this compound would be able to covalently and stably modify the side chain SH of the catalytic cysteine, which would account for the experimentally observed irreversible inhibition of the M^{pro}-catalyzed reaction by this covalent inhibitor.

5.2.3.6 *A Covalent Inhibitor for the Thermotoga maritima α-Galactosidase-catalyzed Reaction*

As described in Section 1.2.3.3, glycoside hydrolases catalyze the hydrolysis of the glycosidic bond (an acetal linkage) in various carbohydrate molecules, and the glycoside hydrolase-catalyzed reaction plays an important regulatory role in various life processes *via* helping to modulate the structure and function of various carbohydrate molecules together with the glycosyltransferase-catalyzed reaction. As shown in Figure 1.18(A) and Figure 5.21(A),

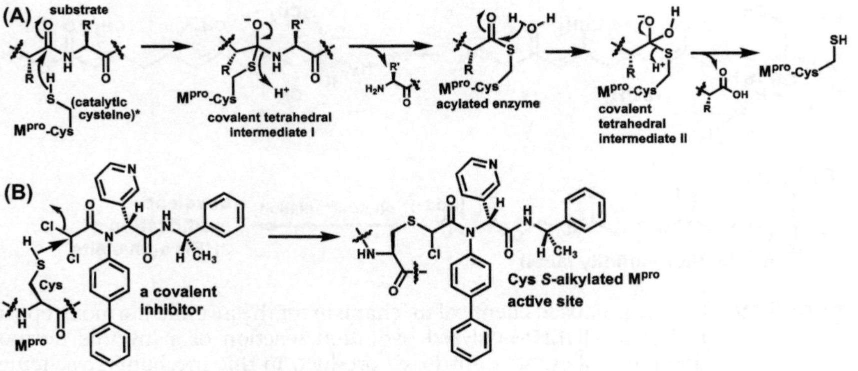

Figure 5.20 (A) The proposed chemical mechanism for the SARS-CoV-2 main protease (M^{pro})-catalyzed amide hydrolysis reaction. R and R′ refer to the side chains flanking the scissile peptide bond. Cys, an active site catalytic cysteine residue. The asterisk in "(catalytic cysteine)*" indicates that its side chain SH would be deprotonated and thus activated by an active site general base (*e.g.* histidine side chain; not depicted) before the nucleophilic attack. (B) The side chain *S*-alkylation of the M^{pro} active site catalytic Cys residue by the indicated α-dichloroacetamide-based structural mimic of the M^{pro} reaction substrate, which would account for the experimentally observed irreversible inhibition of the M^{pro}-catalyzed reaction by the indicated covalent inhibitor.

the α-galactosidase-catalyzed galactosidation reaction entails a double displacement mechanism involving the formation of the hydrolysis-labile covalent enzyme intermediate with an aspartic acid residue *via* the depicted oxonium transition state and the subsequent hydrolysis of this covalent enzyme intermediate liberating the glycone part of the substrate, which would bring about a retention of the α-configuration at the anomeric carbon.[28]

Figure 5.21(B) depicts the *Thermotoga maritima* α-galactosidase (TmGalA)-catalyzed processing of the indicated cyclohexenyl carbasugar-based covalent inhibitor with the formation of the depicted hydrolysis-resistant covalent enzyme intermediate, which would only slowly be hydrolyzed to give rise to the glycone part of the inhibitor and a regenerated TmGalA active site.[28] Due to this slow hydrolysis, the compound would be a covalent inhibitor, yet not an irreversible inhibitor.

5.2.4 Applications with Lyases

In this section, notable examples of the (photo)affinity enzyme labels and covalent inhibitors for enzymatic lyation reactions are elaborated.

5.2.4.1 *An Affinity Label for 3-Hydroxy-3-methylglutaryl Coenzyme A Lyase*

3-Hydroxy-3-methylglutaryl coenzyme A (HMG-CoA) lyase catalyzes the metal cation (*e.g.* Mg^{2+})-dependent splitting of HMG-CoA to afford acetoacetate

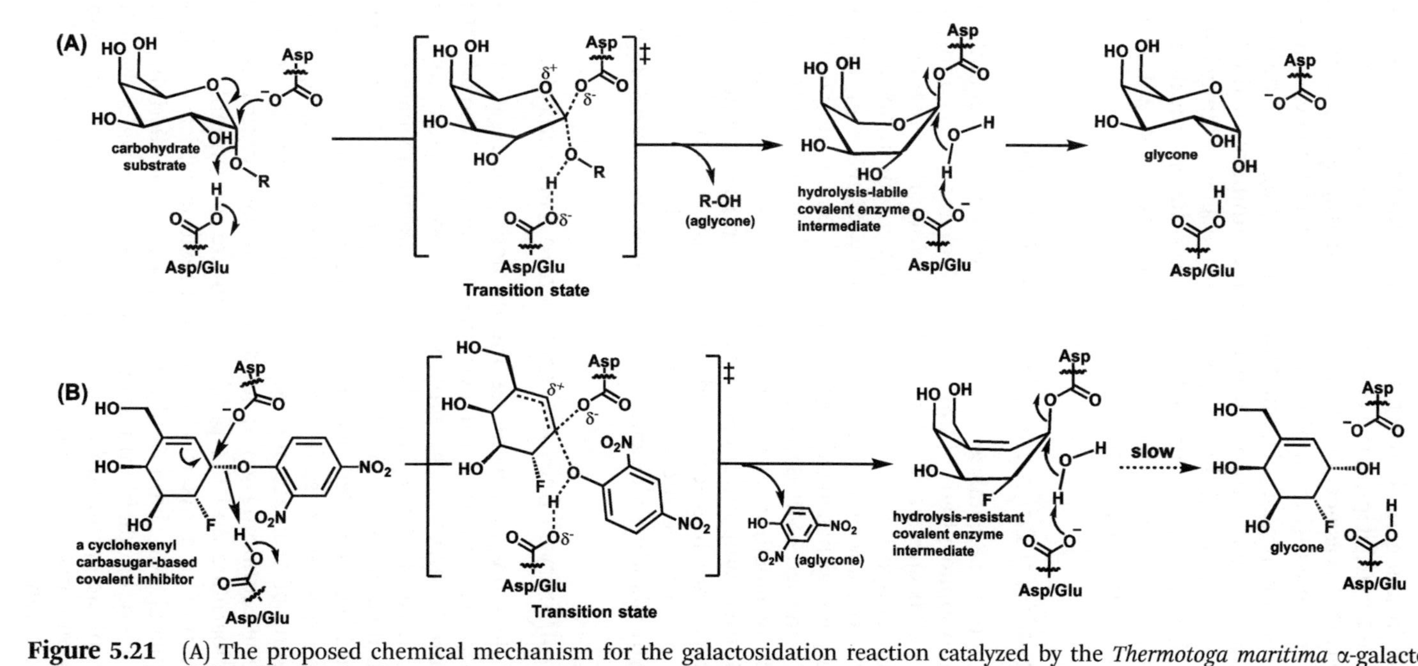

Figure 5.21 (A) The proposed chemical mechanism for the galactosidation reaction catalyzed by the *Thermotoga maritima* α-galactosidase (TmGalA). This mechanistic scheme is identical to that shown above in Figure 1.18(A). (B) The TmGalA-catalyzed processing of the indicated covalent inhibitor with the formation of the depicted hydrolysis-resistant covalent enzyme intermediate, which would only slowly be hydrolyzed to give rise to the glycone part of the inhibitor and a regenerated TmGalA active site.

and acetyl-coenzyme A (acetyl-CoA).[29,30] This enzymatic reaction plays an important role in ketogenesis when the glucose supply becomes limited to humans. Figure 5.22(A) depicts the chemical mechanism of the *Pseudomonas mevalonii* HMG-CoA lyase-catalyzed reaction, illustrating the critical role of an active site cysteine residue in the catalysis.[29,30] Given the essential role of this cysteine residue in catalysis, one HMG-CoA analog (*i.e.* 2-butynoyl-CoA) was tested and was found to be an irreversible inhibitor of the HMG-CoA lyase-catalyzed reaction and therefore would serve as an affinity label of the active site of this enzyme. Figure 5.22(B) depicts the chemical mechanism that could account for the experimentally observed covalent labeling of HMG-CoA lyase active site by butynoyl-CoA.[29]

5.2.4.2 An Affinity Label for Adenylosuccinate Lyase

Adenylosuccinate lyase catalyzes the splitting of adenylosuccinate to form adenosine monophosphate (AMP) and fumarate, as indicated in Figure 5.23(A). This enzymatic reaction plays an important regulatory role in *de novo* purine biosynthesis. Figure 5.23(A) depicts the proposed chemical mechanism for the adenylosuccinate lyase-catalyzed reaction, which would proceed *via* a general base/acid mechanism.[31,32] One analog of the enzymatic reaction product AMP depicted in Figure 5.23(B) was designed as a potential affinity labeling reagent for the active site of adenylosuccinate lyase. Indeed, this compound was found to be an irreversible inhibitor for the adenylosuccinate lyase-catalyzed reaction, presumably *via* a covalent modification of an active site amino acid side chain which would be most likely a catalytic histidine side chain, as indicated in Figure 5.23(B).[31,32]

5.2.4.3 An Affinity Label for Oxaloacetate Decarboxylase

Oxaloacetate decarboxylase catalyzes the divalent metal cation-dependent splitting of oxaloacetate to form pyruvate and carbon dioxide (CO_2), as shown in Figure 5.24(A).[33,34] This enzymatic reaction plays an important role in re-feeding pyruvate back into the central metabolic pathway in both prokaryotic and eukaryotic cells. As shown in Figure 5.24(A), this enzymatic reaction could be regarded as just a simple metal cation-facilitated β-elimination on a β-keto acid. Figure 5.24(B) depicts the S_NAr-mediated covalent labeling of oxaloacetate decarboxylase active site by the indicated dichlorotriazine-containing compound structurally mimicking both the substrate oxaloacetate and the product pyruvate, presumably *via* the side chain alkylation of an active site amino acid residue, which would be most likely the depicted side chain *N*-alkylation of a catalytic histidine residue.[33] It should be noted that dichlorotriazine is able to react covalently with all the known nucleophilic groups on the 20 standard amino acid side chains. As such, this compound was found to inhibit irreversibly the oxaloacetate decarboxylase-catalyzed reaction.

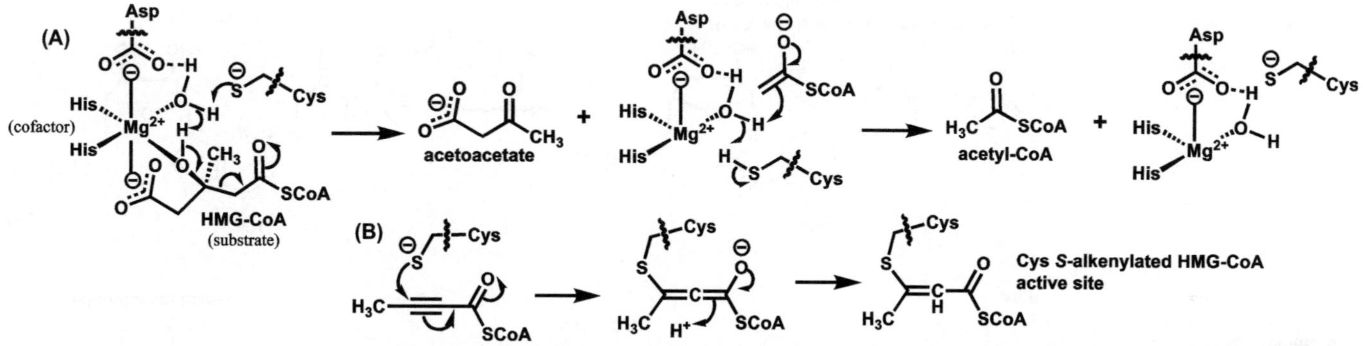

Figure 5.22 (A) The proposed chemical mechanism for the *Pseudomonas mevalonii* HMG-CoA lyase-catalyzed Mg^{2+}-dependent cleavage of HMG-CoA to acetoacetate and acetyl-coenzyme A (acetyl-CoA). It is worth noting that an active site cysteine (Cys) acts as a general base/acid in catalysis. (B) The proposed chemical mechanism that could account for the HMG-CoA active site Cys *S*-alkenylation by 2-butynoyl-CoA.

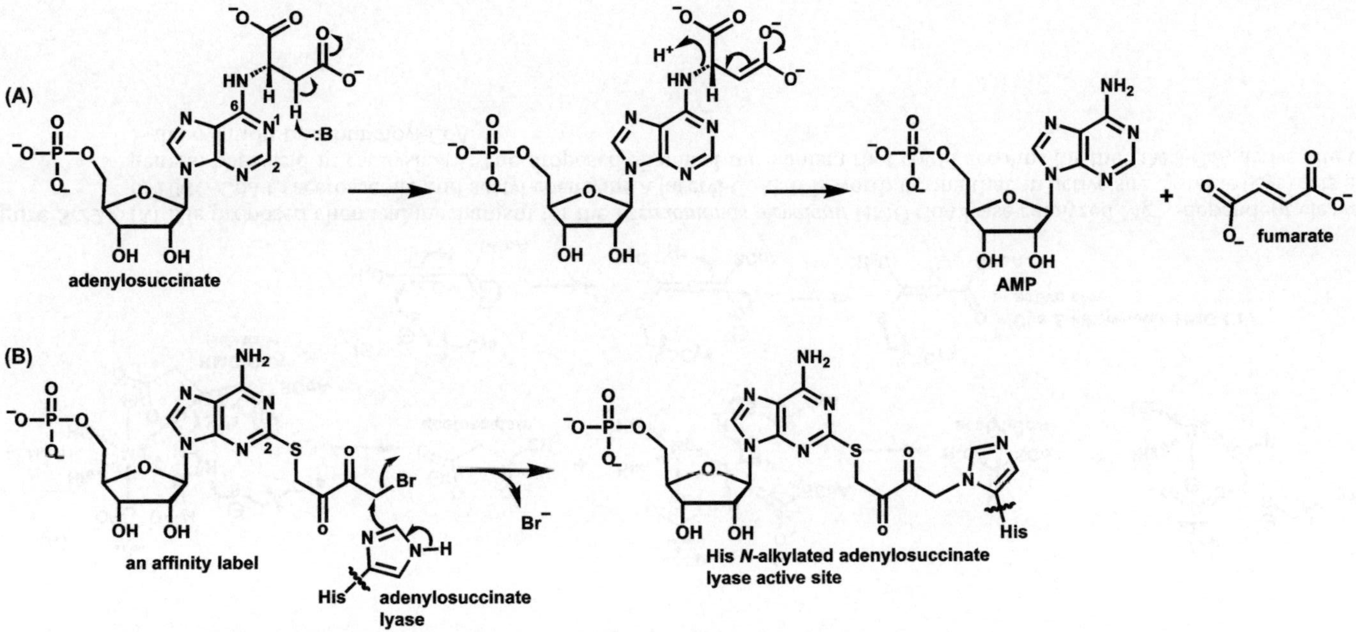

Figure 5.23 (A) The proposed chemical mechanism for the adenylosuccinate lyase-catalyzed cleavage of adenylosuccinate to AMP and fumarate. Note: (i) :B refers to an active site general base which can be a histidine residue or a serine residue; (ii) the N6-protonation is shown here, however, the N1-protonation could be also possible; this proton source could be the bulk solution or an active site general acid. (B) The proposed chemical mechanism that could account for the adenylosuccinate lyase active site histidine (His) side chain *N*-alkylation by the indicated affinity labeling reagent which is an AMP structural analog.

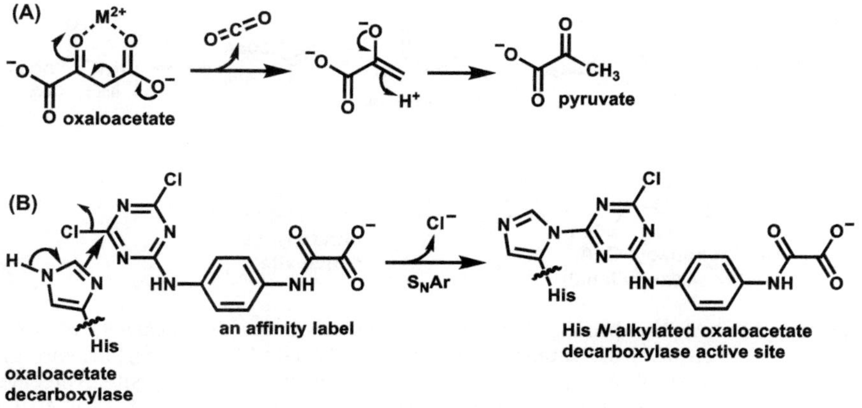

Figure 5.24 (A) The proposed chemical mechanism for the *Pseudomonas stutzeri* oxaloacetate decarboxylase-catalyzed cleavage of oxaloacetate to pyruvate and CO_2. Note: the divalent metal cation M^{2+} could be either Mg^{2+} or Mn^{2+}, which would serve as an electron sink and help to anchor/orient substrate/intermediate/product at enzyme active site. (B) The proposed chemical mechanism that could account for the oxaloacetate decarboxylase active site His side chain *N*-alkylation by the indicated affinity labeling reagent which is a structural analog of oxaloacetate and pyruvate.

5.2.4.4 A Photo-affinity Label for Citrate Lyase

The citrate lyase from *Klebsiella aerogenes* is a multi-enzyme complex composed of subunits α, β, and γ, with subunit γ being an acyl-carrier protein (ACP) with a CoA-like prosthetic group, subunit α being an acyl-transferase catalyzing the formation of (3*S*)-citryl-ACP from citrate, and subunit β being an acyl lyase catalyzing the splitting of the citryl-ACP intermediate to form oxaloacetate and acetyl-ACP.

Figure 5.25(B) depicts the chemical structure of *p*-azidobenzoyl-CoA, which is a structural analog of the reaction substrate and product depicted in Figure 5.25(A),[35] which has been designed as an azide-based photo-affinity label for citrate lyase.[35] Once photolyzed, *p*-azidobenzoyl-CoA was shown to be able to label the β subunit of the citrate lyase multi-enzyme complex, which would suggest that the native substrate (3*S*)-citryl-ACP's binding site on the β subunit is also capable of accommodating an aryl-CoA like *p*-azidobenzoyl-CoA, consistent with the fact that (3*S*)-citryl-CoA also serves as an *in vitro* substrate for the lyase reaction. Presumably, *p*-azidobenzoyl-CoA would be able to covalently modify the β subunit lyase active site in response to photo-irradiation following its binding to the active site in darkness. Under such conditions, the azide functionality would furnish the highly electrophilic nitrene that can be inserted into any proximal covalent bond.

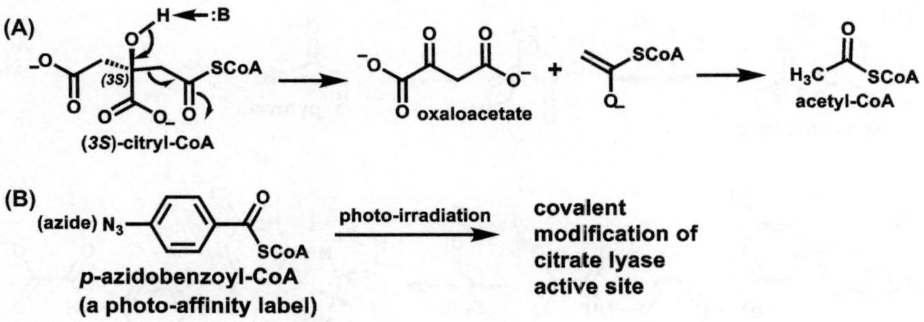

(A)

H←:B ... (3S) ... SCoA → oxaloacetate + SCoA → H_3C SCoA acetyl-CoA

(3S)-citryl-CoA

(B)

(azide) N_3 ... SCoA —photo-irradiation→ covalent modification of citrate lyase active site

p-azidobenzoyl-CoA
(a photo-affinity label)

Figure 5.25 (A) The proposed chemical mechanism for the *Klebsiella aerogenes* citrate lyase-catalyzed cleavage of (3*S*)-citryl-CoA to oxaloacetate and acetyl-CoA. :B, a general base at the citrate lyase active site. Note: even though citrate lyase catalyzes *in vivo* the splitting of citryl-ACP to oxaloacetate and acetyl-ACP, citryl-CoA was also shown to be an *in vitro* substrate for this enzymatic reaction with the formation of oxaloacetate and acetyl-CoA. ACP, acyl-carrier protein, *i.e.* the γ subunit of the citrate lyase multi-enzyme complex with a CoA-like prosthetic group. (B) The chemical structure of *p*-azidobenzoyl-CoA, a structural analog of the reaction substrate and product depicted in (A), which was designed as an azide-based photo-affinity label for citrate lyase.

5.2.4.5 A Covalent Inhibitor for the Isocitrate Lyase-catalyzed Reaction

Isocitrate lyase isoforms 1 and 2 are both able to catalyze the Mg^{2+}-dependent splitting of isocitrate to glyoxylate and succinate, as indicated in Figure 5.26(A).[36] The isocitrate lyase-catalyzed reaction is crucial for the survival and virulence of *Mycobacterium tuberculosis*, and its inhibition would possess an anti-tuberculosis effect.

Figure 5.26(A) also depicts the proposed chemical mechanism for the isocitrate lyase-catalyzed reaction which would follow a general acid/base mechanism.[36] Specifically, it proceeds *via* an initial proton abstraction from the 2-OH of isocitrate by an aspartate side chain carboxylate (general base) through an intervening Mg^{2+}-activated water molecule, which gives rise to glyoxylate and the depicted enolate form of succinate. The subsequent protonation of this enolate structure by a cysteine side chain SH (general acid) would afford the second product of the enzymatic reaction, *i.e.* succinate.

Given the afore-mentioned proposed participation of a cysteine side chain SH as a general acid during the enzymatic catalysis, the natural product itaconate which is a structural analog of the isocitrate lyase reaction product succinate and harbors an α,β-unsaturated Michael acceptor for the nucleophilic thiolate anion (S^-) was tested as a potential covalent inhibitor for the isocitrate lyase-catalyzed reaction. Indeed, itaconate was found to be an irreversible covalent inhibitor.[36] Figure 5.26(B) depicts the proposed mechanism for this inhibitory action, with the key mechanistic event being the Michael addition of the indicated S^- onto the α,β-unsaturated structure of itaconate.

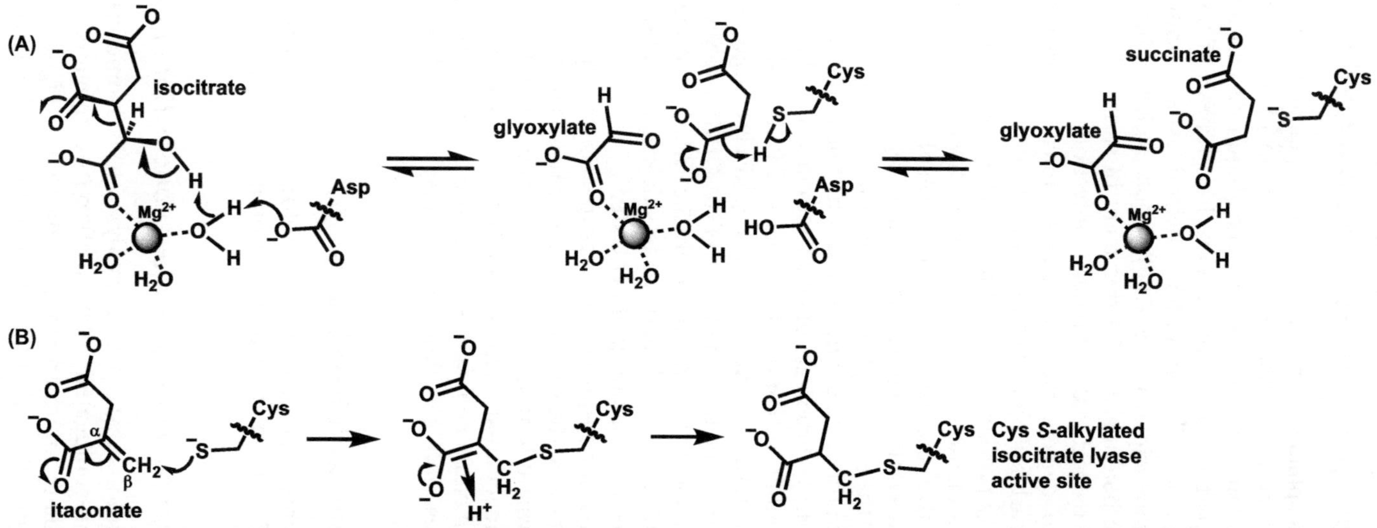

Figure 5.26 (A) The proposed chemical mechanism for the *Mycobacterium tuberculosis* isocitrate lyase-catalyzed Mg^{2+}-dependent splitting of isocitrate to form glyoxylate and succinate. Cys, an active site cysteine residue acting as a general acid during enzymatic catalysis; Asp, an active site aspartate residue acting as a general base during enzymatic catalysis. (B) The side chain *S*-alkylation of the isocitrate lyase active site catalytic Cys residue by itaconate, a structural analog of the isocitrate lyase reaction product succinate, which would account for the experimentally observed irreversible inhibition of the isocitrate lyase-catalyzed reaction by itaconate.

5.2.4.6 *A Covalent Inhibitor for the Orotidine 5′-Monophosphate Decarboxylase-catalyzed Reaction*

Orotidine 5′-monophosphate decarboxylase (ODCase) catalyzes the decarboxylation of orotidine 5′-monophosphate (OMP) to form uridine 5′-monophosphate (UMP), as shown in Figure 5.27(A).[37,38] The ubiquitous ODCase-catalyzed reaction plays an important role in the *de novo* biosynthesis of UMP and the subsequent formation of other pyrimidine nucleotides (*e.g.* uridine triphosphate (UTP) and cytidine triphosphate (CTP)), whose activity inhibition would possess a therapeutic potential for the treatment of various human diseases such as those caused by bacterial and parasite infections.[38]

Figure 5.27(A) also depicts the proposed chemical mechanism for the ODCase-catalyzed decarboxylation of OMP to form UMP.[38] From the standpoint of reaction free energy profile, the enzymatic strategy is combined substrate destabilization and transition state stabilization. Specifically, the judicious positioning of an active site aspartate residue (Asp) would destabilize the ground state substrate OMP due to an electrostatic repulsion between its side chain carboxylate and the 6-carboxylate of OMP. However, this charge repulsion would be weakened in the depicted transition state due to the negative charge de-localization from the 6-carboxyate of OMP onto its C-6 carbon. In addition, the side chain ammonium cation of a proximal active site lysine residue (Lys) would incur an electrostatic attraction with the negatively charged 6-carboxylate and C-6 of OMP. As such, the transition state of the ODCase-catalyzed reaction would be stabilized. As also depicted in Figure 5.27(A), the subsequent protonation of C-6 of OMP by the indicated lysine side chain ammonium cation would help to produce the reaction product UMP.

Given the afore-mentioned proposed participation of a lysine side chain during enzymatic catalysis, the UMP analog 6-I-UMP was tested and found to be an irreversible covalent inhibitor for the ODCase-catalyzed reaction.[37] The proposed chemical mechanism depicted in Figure 5.27(B) *via* Michael addition and elimination could be responsible for the covalent inhibition of the ODCase-catalyzed reaction *via* side chain *N*-alkylation of an active site catalytic Lys residue.

5.2.5 Applications with Isomerases

In this section, notable examples of the (photo)affinity enzyme labels and covalent inhibitors for enzymatic isomerization and racemization reactions are elaborated.

5.2.5.1 *An Affinity Label for 4-Oxalocrotonate Tautomerase*

4-Oxalocrotonate tautomerase (4-OT) catalyzes the isomerization of 4-oxalocrotonate (an un-conjugated α-keto acid) to its conjugated isomer *via* a fully conjugated enolate intermediate, as indicated in Figure 5.28(A).[39] As also indicated in Figure 5.28(A), 4-OT employs a general acid/base catalytic strategy to achieve the highly efficient rate acceleration, with its N-terminal

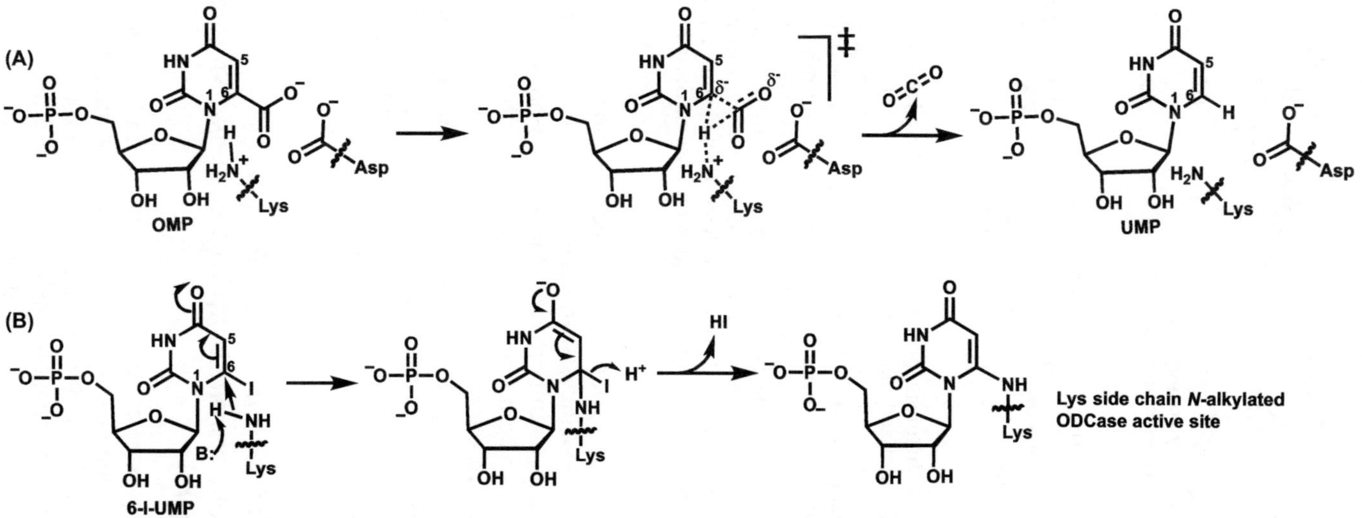

Figure 5.27 (A) The proposed chemical mechanism *via* the indicated transition state structure for the *Bacillus subtilis* orotidine 5′-monophosphate decarboxylase (ODCase)-catalyzed decarboxylation of orotidine 5′-monophosphate (OMP) to form uridine 5′-monophosphate (UMP). Asp, an active site aspartate residue whose positioning at ODCase active site would destabilize the ground state substrate OMP *via* an electrostatic repulsion between its side chain carboxylate anion and the 6-carboxylate of OMP; Lys, an active site lysine residue whose ammonium side chain would help to stabilize the reaction transition state *via* electrostatic attraction with the negatively charged 6-carboxylate and C-6 of OMP, and to protonate the C-6 of OMP *en route* to the reaction product UMP. Of note, the negative charge delocalization onto C-6 from the 6-carboxyate of OMP would also help to stabilize the transition state *via* a reduced electrostatic repulsion with the negatively charged side chain carboxylate of the Asp residue. (B) The side chain *N*-alkylation of the ODCase active site catalytic Lys residue by 6-I-UMP, a structural analog of the ODCase reaction product UMP, which would account for the experimentally observed irreversible inhibition of the ODCase-catalyzed reaction by 6-I-UMP. B:, a general base at the ODCase active site.

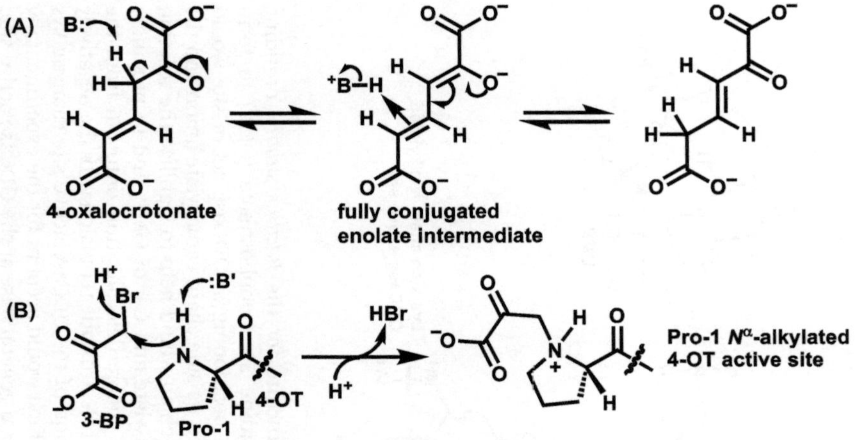

Figure 5.28 (A) The proposed chemical mechanism for the 4-oxalocrotonate tautomerase (4-OT)-catalyzed isomerization of the un-conjugated α-keto acid 4-oxalocrotonate to its conjugated isomer *via* the depicted fully conjugated enolate intermediate. B: refers to a general base during this single-base enzymatic catalysis, which could be the N-terminal proline residue (Pro-1) α-amino group. (B) The N^{α}-alkylation of the 4-OT active site catalytic Pro-1 residue by 3-bromopyruvate (3-BP), a structural analog of the 4-OT reaction substrate 4-oxalocrotonate, which would account for the experimentally observed irreversible inhibition of the 4-OT-catalyzed reaction by 3-BP. :B′, a general base at the 4-OT active site.

proline residue (Pro-1) α-amino group having been proposed to be the catalytic base in the single-base catalysis. In order to help to elucidate the active site of 4-OT and the catalytic importance of its Pro-1, 3-bromopyruvate (3-BP) was tested and found to be a covalent irreversible inhibitor for the 4-OT-catalzyed isomerization reaction.

Figure 5.28(B) depicts the nucleophilic substitution reaction between 3-BP and the α-amino group of the Pro-1 residue of 4-OT leading to the covalent modification of Pro-1 and the 4-OT active site, which could account for the experimentally observed covalent irreversible inhibition of the 4-OT-catalyzed reaction by 3-BP.[39]

5.2.5.2 An Affinity Label for the Human Placental 3β-Hydroxy-5-ene-steroid Dehydrogenase and Steroid 5→4-ene-isomerase

Human placental 3β-hydroxy-5-ene-steroid dehydrogenase and steroid 5→4-ene-isomerase are the two different enzymatic activities dwelling in a single protein, but supported by two different active sites. Figure 5.29(A) shows one example of the enzymatic transformation catalyzed by these two enzymatic activities, *i.e.* the conversion of dehydroepiandrosterone to androstenedione *via* the depicted 3-keto-5-ene-steroid intermediate, which is catalyzed by the

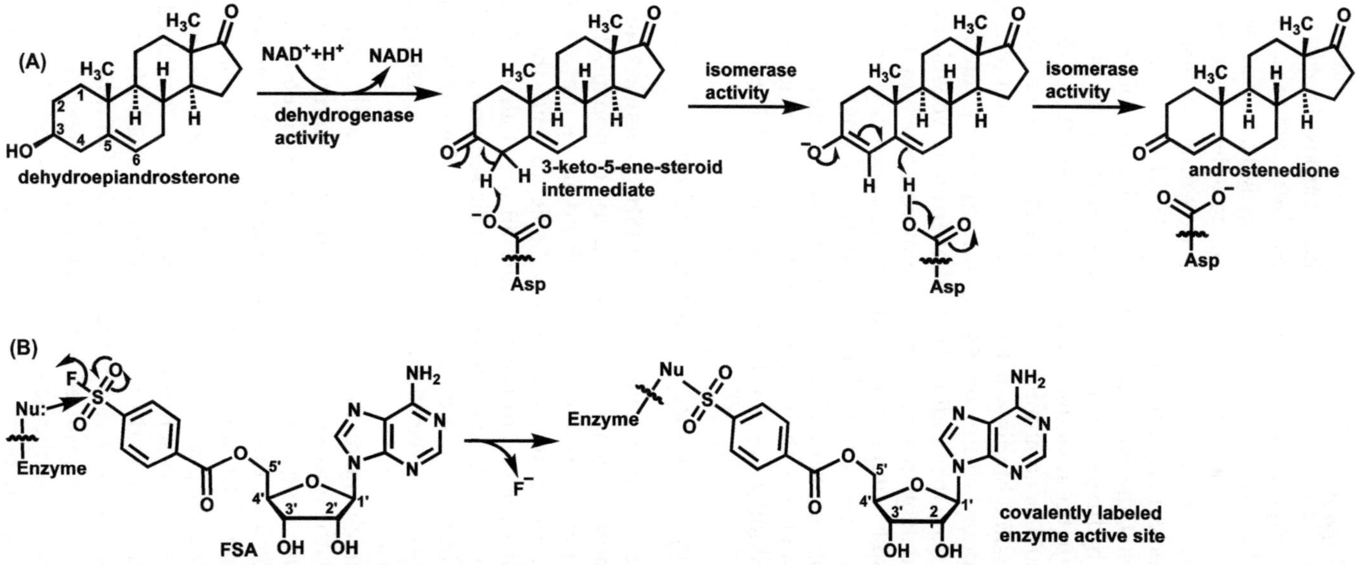

Figure 5.29 (A) The transformation of dehydroepiandrosterone to androstenedione *via* the depicted 3-keto-5-ene-steroid intermediate which are catalyzed in tandem by the human placental 3β-hydroxy-5-ene-steroid dehydrogenase and steroid 5→4-ene-isomerase, and the proposed chemical mechanism for the latter activity. Asp, aspartic acid. Please see other figures (*e.g.* Figure 3.7) in this book for the full chemical structures of NAD^+ and NADH. (B) The covalent labeling of the coenzyme binding site of "Enzyme" *via* sulfonylation of a nucleophile (Nu:) by 5′-[*p*-(fluorosulfonyl)benzoyl]adenosine (FSA). Note: (i) "Enzyme" refers to the protein harboring the human placental 3β-hydroxy-5-ene-steroid dehydrogenase and steroid 5→4-ene-isomerase activities; (ii) Nu: could be most likely a side chain nucleophilic group of an amino acid residue in the coenzyme binding site.

afore-mentioned dehydrogenase and isomerase enzymatic activities in tandem.[40] As depicted in Figure 5.29(A), the isomerase activity entails a general acid/base chemical mechanism with the side chain carboxylic acid and carboxylate of an active site aspartic acid residue respectively serving as the general acid and the general base during the enzymatic catalysis.[41] Because the isomerase activity is dehydrogenase-linked, its enzymatic activity could be affected by the coenzyme NAD^+-derived product of the dehydrogenase-catalyzed reaction, *i.e.* NADH. To help analyze the coenzyme NAD^+/NADH binding site and binding behavior during the enzymatic catalysis by human placental 3β-hydroxy-5-ene-steroid dehydrogenase and steroid 5→4-ene-isomerase, the NADH structural analog 5′-[*p*-(fluorosulfonyl)benzoyl]adenosine (FSA) depicted in Figure 5.29(B) was tested as an affinity labeling reagent.[40] It could be expected that, following a reversible binding of FSA to the coenzyme binding site, the electrophilic fluorosulfonyl group on FSA would covalently interact with a proximal nucleophile from the binding site leading to an irreversible inhibition of the afore-mentioned dehydrogenase and/or isomerase activities. Indeed, FSA was found to irreversibly inhibit both of these two enzymatic activities with comparable efficiency. This observation suggests that a single and common coenzyme binding site is involved in the catalysis, not just by the dehydrogenase activity, but also by the isomerase activity, and further implies that the steroid substrate binding sites for the two enzymatic activities are different.

5.2.5.3 A Photo-affinity Label for Squalene:Hopene Cyclase

The bacterial enzyme squalene:hopene cyclase (SHC) catalyzes an electrophile-initiated cyclization of squalene to hop-22-ene or hopan-22-ol *via* the depicted 22-carbocation intermediate, as depicted in Figure 5.30(A).[42] As depicted, amazingly, the SHC active site is able to fold the extended hydrocarbon chain of squalene to the depicted all pre-chair conformation. The ultimate proton (H^+) source in this mechanistic scheme could be more that from a SHC active site amino acid side chain than that from bulk aqueous solution, since this electrophile-initiated electron pushing cascade is mechanistically specific.

In order to further elucidate the active site of SHC with respect to its binding interaction with substrate molecules, inhibitory compounds like the one depicted in Figure 5.30(B) were identified. This example behaves as a very potent non-terpenoid inhibitor for the *Alicyclobacillus acidocaldarius* SHC-catalyzed reaction.[42] Of note, the alkyl-ammonium cation in this inhibitory compound would possibly mimic the 22-carbocation depicted in Figure 5.30(A). Due to the presence in its structure of a substituted benzophenone moiety, this compound was exploited for a photo-affinity labeling experiment. It was fortuitously found that its use in dark and then under photo-irradiation conditions resulted in a covalent labeling of the SHC active site,[42] presumably because of the benzophenone-derived highly electrophilic carbene under photo-irradiation conditions that can be inserted into any proximal covalent bond.

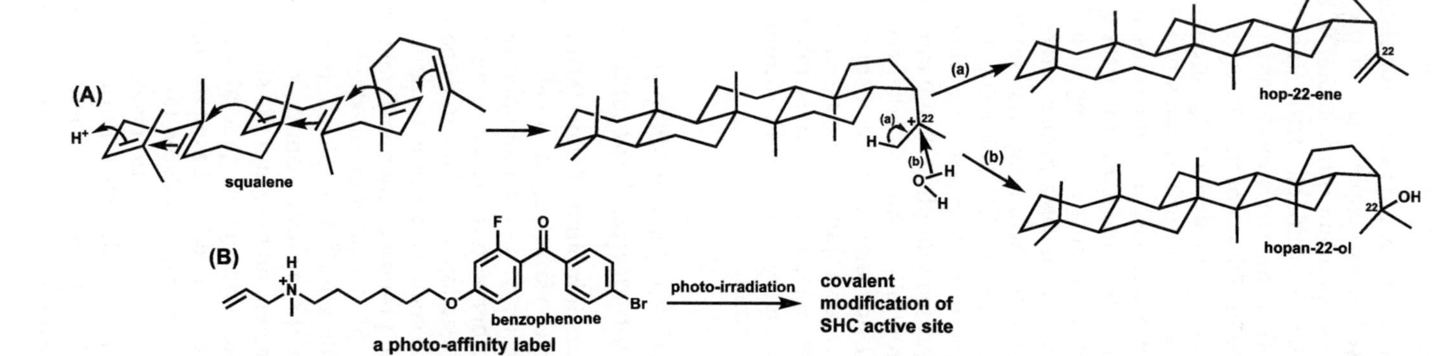

Figure 5.30 (A) The proposed chemical mechanism for the squalene:hopene cyclase (SHC)-catalyzed electrophile-initiated cyclization of squalene to hop-22-ene or hopan-22-ol *via* the depicted 22-carbocation intermediate. It is amazing that the SHC active site is able to fold the extended hydrocarbon chain of squalene to the depicted all pre-chair conformation. The ultimate proton (H^+) source during this enzymatic catalysis could be a SHC active site amino acid side chain. (B) The chemical structure of a benzophenone-based photo-affinity label for SHC active site.

5.2.5.4 A Photo-affinity Label for DNA Gyrase

The tetrameric bacterial enzyme DNA gyrase, a topoisomerase II (Topo II) family member, is composed of two copies each of the subunits GyrA and GyrB, and catalyzes the ATP-dependent introduction of negative super-coils into bacterial circular double-stranded DNA (dsDNA), as illustrated schematically in Figure 5.31(A) for the strand passage mechanism of this enzymatic reaction.[43–45] The inhibition of the DNA gyrase-catalyzed reaction has been regarded as a strategy for developing anti-bacterials such as the broad-spectrum bacteriocidal agent norfloxacin, since this inhibitory action would disturb the DNA-templated cellular processes including replication and transcription. In order to help elucidate the DNA gyrase active site and the specific interactions involved in the binding of a quinolone-class anti-bacterial such as norfloxacin in the DNA gyrase active site, the azide-based photo-affinity labeling reagent that was built upon norfloxacin and is shown in Figure 5.31(B) has been designed.[43] Following the incubation of this compound with the tetrameric DNA gyrase in darkness together with the dsDNA substrate and the subsequent photo-irradiation, this compound was found to be able to preferentially covalently label the GyrA subunit of the tetrameric enzyme, which lends another piece of support for the notion that the direct interaction target for a quinolone-class anti-bacterial is the GyrA subunit of DNA gyrase. Of note, under photo-irradiation the azide functionality would furnish the highly electrophilic nitrene that can be inserted into any proximal covalent bond.

5.2.5.5 A Covalent Inhibitor for the Macrophage Migration Inhibitory Factor-catalyzed Tautomerization Reaction

The macrophage migration inhibitory factor (MIF) is a cytokine which can exert its functional roles *via* acting on a cell surface receptor, it also possesses an enzymatic activity, *i.e.* a tautomerase activity as indicated in Figure 5.32(A).[46–48] Together with the receptor-mediated activity, the MIF-catalyzed tautomerization reaction also plays an important pathological role in various human diseases, such as the inflammatory diseases and cancer.

Figure 5.32(A) shows a proposed chemical mechanism for the MIF-catalyzed tautomerization of the *in vitro* substrate D-dopachrome to 5,6-dihydroxyindole-2-carboxylic acid.[47] As depicted, this is a general acid/base catalytic process in which the neutral α-amino of the N-terminal proline (Pro-1) of MIF serves as the general base and the side chain ammonium of a catalytic lysine (Lys) residue serves as the general acid; general base B: may be an activated catalytic water molecule. Based on this, a hydroxyquinoline-based covalent inhibitor of the MIF-catalyzed tautomerization reaction was designed and whose proposed mode of action is depicted in Figure 5.32(B).[46] The resultant formation of the depicted Pro-1 N^{α}-alkylated MIF would presumably be responsible for the experimentally observed irreversible inhibition of the MIF-catalyzed tautomerization reaction.

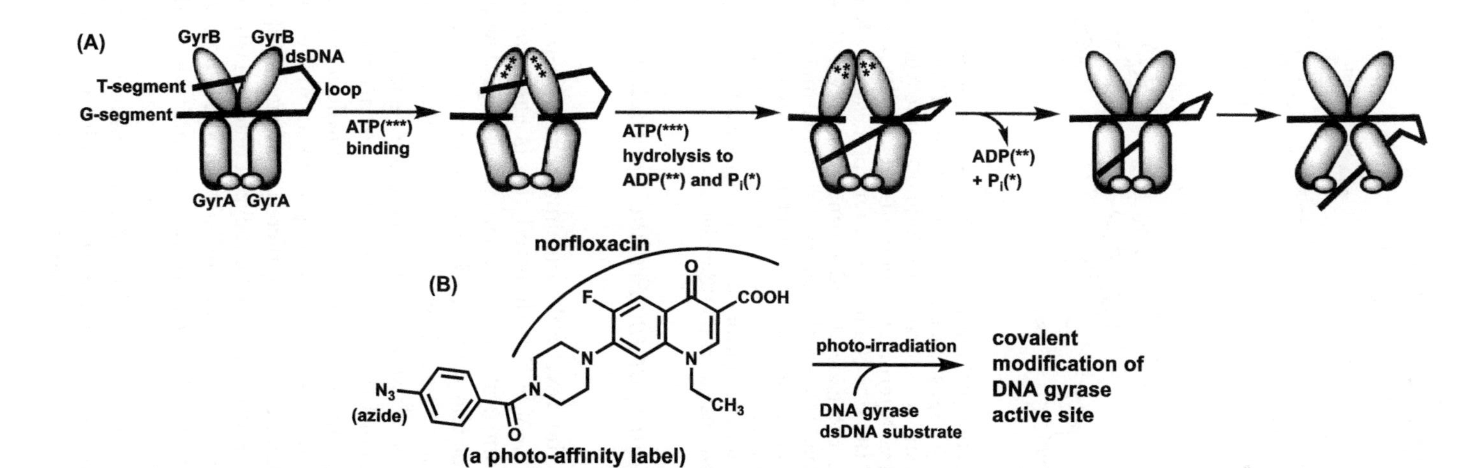

Figure 5.31 (A) A schematic drawing illustrating the ATP-dependent strand passage mechanism for the tetrameric bacterial DNA gyrase-catalyzed reaction leading to the introduction of negative supercoil into bacterial circular double-stranded DNA (dsDNA). Specifically, ATP binding onto the GyrB subunit would promote the dimerization of the two GyrB subunits *via* their ATP binding sites, the cleavage (DNA backbone phosphodiester breakage) of the G-segment of the circular dsDNA substrate, and the passage of the T-segment of the circular dsDNA substrate through the gap on the G-segment; ATP hydrolysis to ADP and inorganic phosphate (P_i) would promote the re-ligation (DNA backbone phosphodiester re-formation) of the gap on the G-segment and the release of the T-segment from the holoenzyme. Note: only G- and T-segments and the loop between them of the circular dsDNA substrate are shown in this illustration. (B) The chemical structure of an azide-based photo-affinity label for DNA gyrase active site. Also indicated is the chemical structure of the broad-spectrum bacteriocidal agent norfloxacin, an inhibitor of the topoisomerase II (Top II)-catalyzed reaction.

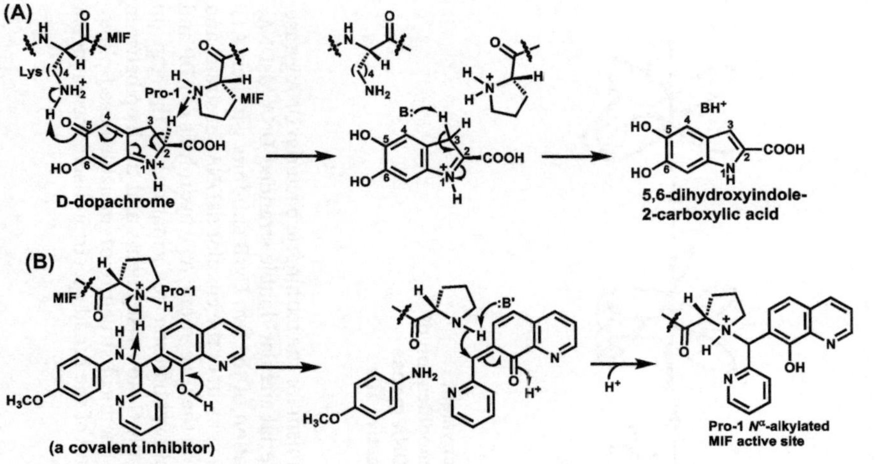

Figure 5.32 (A) The proposed chemical mechanism for the macrophage migration inhibitory factor (MIF)-catalyzed tautomerization of the *in vitro* substrate D-dopachrome to 5,6-dihydroxyindole-2-carboxylic acid. While the α-amino of the N-terminal proline (Pro-1) of MIF serves as the general base and the side chain ammonium of a catalytic lysine (Lys) residue serves as the general acid during this enzymatic catalysis, general base B: may be an activated catalytic water molecule. (B) The proposed mode of action of the depicted hydroxyquinoline-based covalent inhibitor of the MIF-catalyzed tautomerization reaction, which would result in the formation of the depicted Pro-1 N^{α}-alkylated MIF. :B′, a general base at the MIF active site.

5.2.5.6 *A Covalent Inhibitor for the Proline Racemase-catalyzed Reaction*

Proline racemase catalyzes the reversible racemization of L-proline and D-proline, as shown in Figure 5.33(A).[49] As depicted, this is a general acid/base enzymatic catalysis in which one cysteine side chain thiolate (S^-) would serve as the general base and the side chain thiol (SH) of another active site cysteine would serve as the general acid. Based on this mechanism, especially the engagement of the side chains of two active site cysteine residues, the compound depicted in Figure 5.33(B) that harbors a Michael acceptor was designed and was observed experimentally to serve as a covalent inhibitor for the proline racemase-catalyzed reaction.[49] The proposed mode of action (Michael addition) of this covalent inhibitor depicted in Figure 5.33(B) for the *Trypanosoma cruzi* proline racemase (*Tc*PRAC)-catalyzed reaction which would result in the formation of the depicted Cys300 *S*-alkylated *Tc*PRAC could be responsible for the experimentally observed irreversible inhibition. Given the importance of the *Tc*PRAC-catalyzed reaction in conferring parasite's immune escape and persistence in a host, the inhibitors for the *Tc*PRAC-catalyzed reaction could be developed into novel therapeutic agents for Chagas disease, which is caused by *Trypanosoma cruzi* infection in humans.

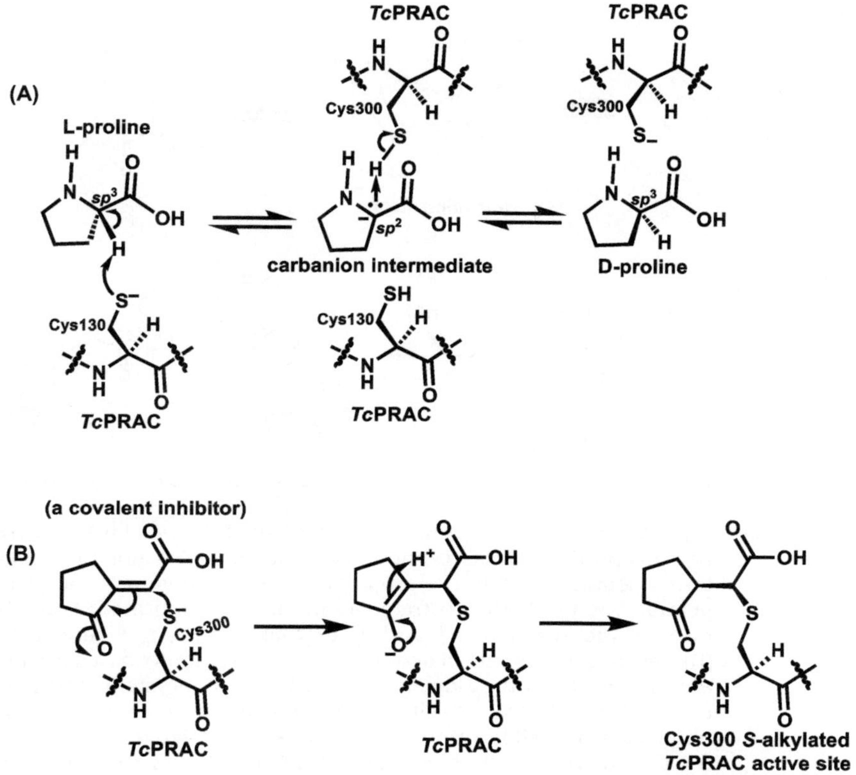

Figure 5.33 (A) The proposed chemical mechanism for the *Trypanosoma cruzi* proline racemase (*Tc*PRAC)-catalyzed reversible racemization of L/D-proline. In the mechanistic scheme for the racemization of L-proline to D-proline, the side chain thiolate (S^-) of the active site cysteine 130 (Cys130) serves as the general base and the side chain thiol (SH) of the active site cysteine 300 (Cys300) serves as the general acid. Note that the hybridization status changes for the stereogenic carbon in the reaction substrate, product, and the carbanion intermediate. (B) The proposed mode of action (Michael addition) of the depicted covalent inhibitor for the *Tc*PRAC-catalyzed racemization reaction, which would result in the formation of the depicted Cys300 *S*-alkylated *Tc*PRAC.

5.2.5.7 A Covalent Inhibitor for the Pin1-catalyzed Isomerization Reaction

As described in Section 4.2.5.1, the peptidyl-prolyl isomerase Pin1 catalyzes the *cis–trans* isomerization of the phosphoSer/Thr-Pro peptide bond in a substrate and this enzymatic isomerization reaction plays a role in regulating cell cycle progression, and therefore its inhibition would possess a potential anti-cancer effect. As depicted in Figure 5.34(A) and Figure 4.8(A), the most salient mechanistic feature of the Pin1-catalyzed reaction is the substrate conformational distortion at the reaction transition state, which

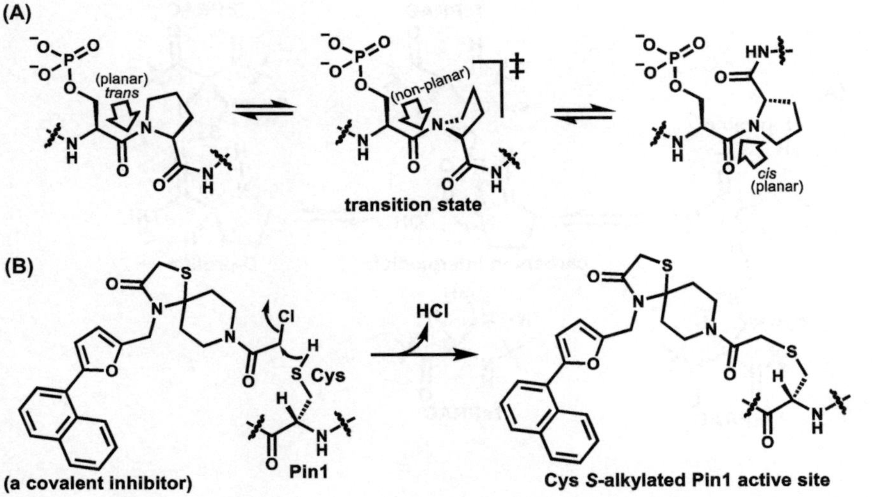

Figure 5.34 (A) The proposed chemical mechanism for the Pin1-catalyzed peptidyl-prolyl *cis–trans* isomerization reaction. The non-planar distorted peptide bond in the proposed transition state and the planar peptide bond in both substrate and product are indicated. Note: this mechanistic scheme is identical to that shown in Figure 4.8(A). (B) The proposed mode of action of a chloroacetamide-based covalent Pin1 inhibitor *via* the side chain *S*-alkylation of a cysteine residue at the Pin1 active site, which would account for the experimentally observed irreversible inhibition of the Pin1-catalyzed isomerization reaction by this inhibitory compound. Note: this covalent inhibitor was discovered by structure-guided optimization of a hit derived from a compound library screening campaign.

actually served as the key rationale for the design of the transition state analog non-covalent inhibitor for the Pin1-catalyzed isomerization reaction described in Section 4.2.5.1 and depicted in Figure 4.8(B). In the work described in this section, the chloroacetamide-based covalent inhibitor depicted in Figure 5.34(B) was discovered by a structure-guided optimization of a hit inhibitor which had been derived from a compound library screening campaign.[50] This inhibitory compound was found experimentally to be an irreversible inhibitor for the Pin1-catalyzed reaction and able to alkylate the side chain SH of a Pin1 active site cysteine residue. The proposed mode of action of this chloroacetamide-based covalent Pin1 inhibitor *via* the side chain *S*-alkylation of a cysteine residue at the Pin1 active site could be responsible for the experimentally observed irreversible inhibition of the Pin1-catalyzed isomerization reaction by this inhibitory compound.

5.2.6 Applications with Ligases

In this section, examples of the (photo)affinity enzyme labels and covalent inhibitors for a few enzymatic ligation reactions are elaborated. It should be noted that few relevant reports have been published, and the examples covered here constitute all those currently available in the literature.

5.2.6.1 An Affinity Label for Isoleucine-tRNA Ligase

Isoleucine-tRNA ligase catalyzes the ATP-dependent synthesis of isoleucyl-tRNA *via* the isoleucyl-AMP intermediate, as shown in Figure 5.35(A).[51,52] The driving force for the formation of the phosphoanhydride-containing high-energy intermediate is the pyrophosphatase (PPase)-catalyzed hydrolysis of one molecule of pyrophosphate (PPi) to two molecules of inorganic phosphate (Pi). It should be noted that the product isoleucyl-tRNA of the isoleucine-tRNA ligase reaction will be next loaded at the A-site of the ribosome and used in the ribosomal peptidyl transferase-catalyzed peptide bond formation reaction as described in Section 4.2.1.3 and Figure 4.3.

In order to facilitate a structure/function analysis of the isoleucine-tRNA ligase active site, a structural analog of isoleucine, *i.e.* isoleucyl-bromomethyl ketone depicted in Figure 5.35(B), was tested as a potential affinity labeling reagent. This compound was found to be a fairly effective affinity label for the isoleucine-tRNA ligase active site and the mechanistic scheme depicted in Figure 5.35(B) could be that responsible for the isoleucine-tRNA ligase active site labeling by isoleucyl-bromomethyl ketone *via* the depicted side chain *S*-alkylation of an active site cysteine residue.[51] Of note, it was also observed that isoleucyl-chloromethyl ketone was unable to label isoleucine-tRNA ligase under the same experimental condition, which would be consistent with that depicted in Figure 5.35(B), in that the enzyme labeling reaction follows a typical S_N2-type nucleophilic substitution mechanism.

5.2.6.2 A Photo-affinity Label for D-Alanyl-D-alanine Ligase

D-Alanyl-D-alanine ligase catalyzes the ATP-dependent peptide bond formation between two molecules of D-alanine to form dipeptide D-alanyl-D-alanine, as shown in Figure 5.36(A).[53,54] Since D-alanyl-D-alanine is a key component of the peptidoglycan layer of the bacterial cell wall, the inhibition of this enzymatic reaction could be a therapeutic strategy for developing novel anti-bacterial agents. As indicated in Figure 5.36(A), the key mechanistic step for the D-alanyl-D-alanine ligase-catalyzed reaction is the α-carboxyl activation of the first D-alanine substrate molecule by ATP in the form of the phosphoanhydride-containing high-energy D-alanyl-phosphate intermediate.[54] Given the dependence of this enzymatic reaction on ATP, a close structural analog of ATP, *i.e.* 8-azido-ATP depicted in Figure 5.36(B), was tested as a potential photo-affinity label for D-alanyl-D-alanine ligase.[53] 8-Azido-ATP was observed to be a photo-affinity labeling reagent for D-alanyl-D-alanine ligase, which would suggest that, following the binding of 8-azido-ATP at the enzyme active site in darkness and then photo-irradiation, the azide functionality of 8-azido-ATP would furnish the highly electrophilic nitrene that can be inserted into any proximal covalent bond, leading to a

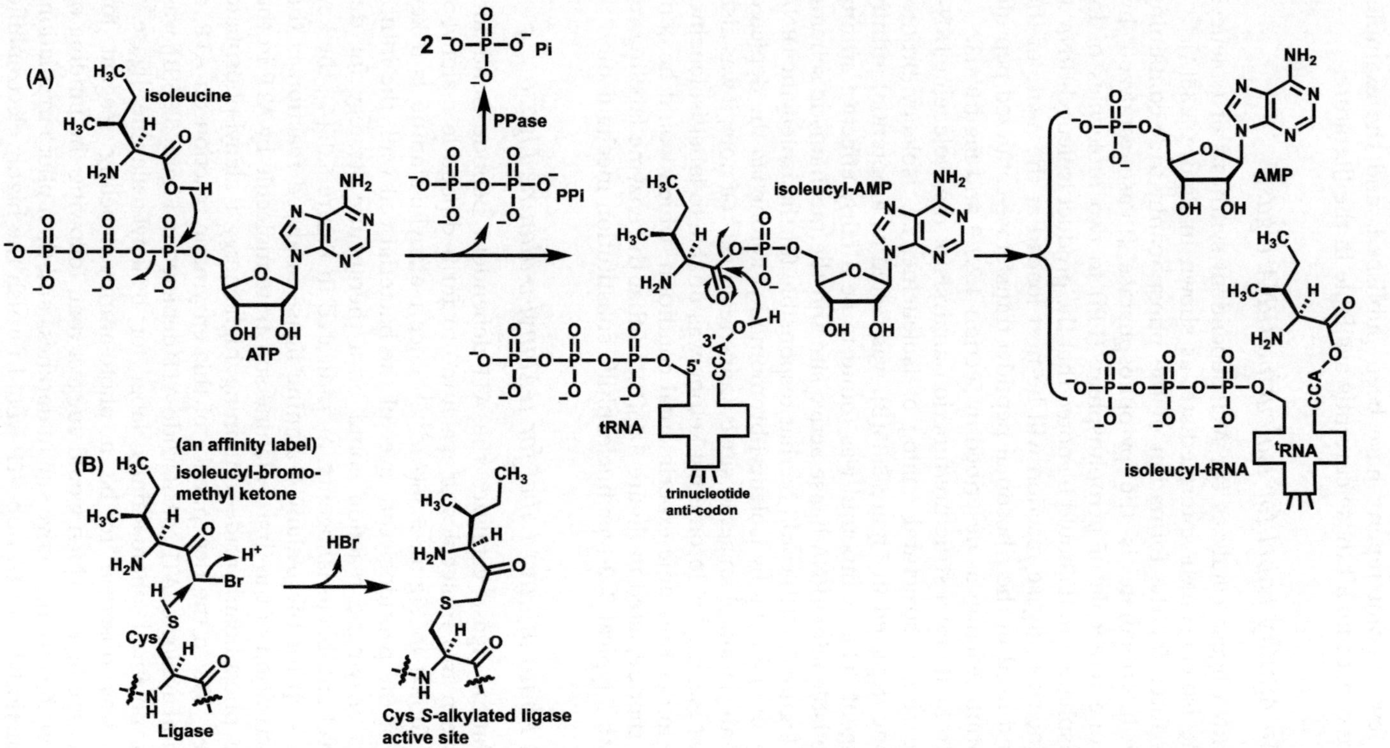

Figure 5.35 (A) The proposed chemical mechanism for the *Escherichia coli* isoleucine-tRNA ligase-catalyzed ATP-dependent synthesis of isoleucyl-tRNA *via* the high-energy phosphoanhydride-containing isoleucyl-AMP intermediate whose formation is driven by the pyrophosphatase (PPase)-catalyzed hydrolysis of pyrophosphate (PPi) to inorganic phosphate (Pi). CCA, the 3′-terminal trinucleotide conserved in tRNA. (B) The proposed chemical mechanism for the isoleucine-tRNA ligase (*i.e.* "ligase") active site Cys *S*-alkylation by the indicated affinity labeling reagent isoleucyl-bromomethyl ketone, a structural analog of isoleucine.

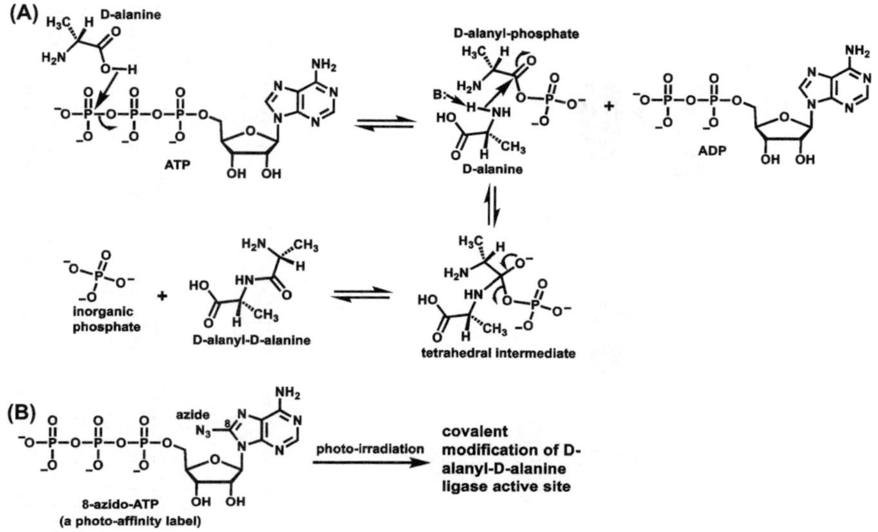

Figure 5.36 (A) The proposed chemical mechanism for the D-alanyl-D-alanine ligase-catalyzed ATP-dependent synthesis of dipeptide D-alanyl-D-alanine, which is a key component of the peptidoglycan layer of the bacterial cell wall. The key mechanistic step is the activation of the α-carboxyl of the first D-alanine substrate molecule by ATP in the form of the phosphoanhydride-containing high-energy D-alanyl-phosphate. B:, a general base at the active site of D-alanyl-D-alanine ligase. (B) The chemical structure of a close structural analog of ATP, *i.e.* 8-azido-ATP, which was tested as an azide-based photo-affinity label for D-alanyl-D-alanine ligase.

covalent modification of the active site of D-alanyl-D-alanine ligase, as schematically depicted in Figure 5.36(B). The availability of such a photo-affinity labeling reagent would promote the analysis of the active site of D-alanyl-D-alanine ligase, being employed to pinpoint the ATP binding site.

5.2.6.3 A Covalent Inhibitor for the RBR E3 Ubiquitin Ligase HOIP-catalyzed Reaction

Protein poly-ubiquitination is an important type of post-translational modification that plays an important role in regulating cellular signaling and protein degradation. Linear ubiquitin chain assembly complex (LUBAC) mediates the synthesis of a particular type of the poly-ubiquitin chain known as the linear poly-ubiquitin chain which is put together on the α-amino of the first methionine residue (*i.e.* Met-1) of a substrate protein with the ubiquitin (Ub) units being inter-connected *via* an isopeptide bond between the C-terminal carboxyl of an incoming Ub unit and an amino group of a Ub unit in an elongating poly-ubiquitin chain. It should be noted that the catalytic subunit of LUBAC is HOIP, which belongs to the ring-between-ring (RBR) family of E3 ubiquitin ligases. Therefore, the inhibition of the

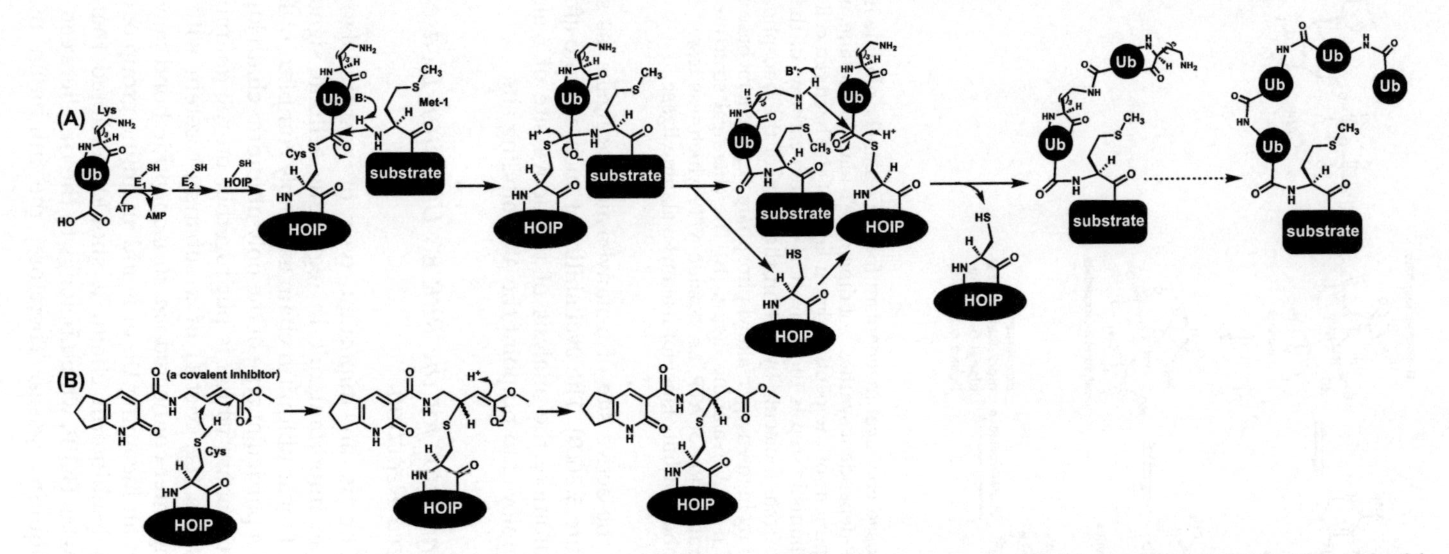

Figure 5.37 (A) The proposed chemical mechanism for the RBR E3 ubiquitin ligase HOIP-catalyzed formation of a linear poly-ubiquitin chain on the α-amino of the first methionine residue (*i.e.* Met-1) of a substrate protein. Ub, ubiquitin; Met, methionine; E_1, the ubiquitin-activating enzyme; E_2, ubiquitin-conjugating enzyme. Note: (i) the SH on E_1, E_2, and HOIP denotes the catalytic cysteine side chain SH group on the respective protein molecules; (ii) B: and B′:, general bases at HOIP active site; (iii) the final enzymatic product shown here harbors four Ub units inter-connected *via* an isopeptide bond between the C-terminal carboxyl of an incoming Ub unit and a lysine ε-amino group in a Ub unit in an elongating poly-ubiquitin chain despite the possible presence of other numbers of the Ub units than four in some substrate proteins modified by poly-ubiquitination. (B) The side chain *S*-alkylation of a HOIP active site catalytic cysteine (Cys) residue by the depicted covalent inhibitor that was identified *via* the fragment-based covalent ligand screening, which could account for the experimentally observed irreversible inhibition of the HOIP-catalyzed reaction by this Michael acceptor-type inhibitory compound.

HOIP-catalyzed reaction could furnish lead compounds for developing novel therapeutic agents for human diseases and chemical tools for a chemical biology-type further mechanistic study on the protein poly-ubiquitination and the corresponding deubiquitination process since linear poly-ubiquitin chains have been known to play a vital role in regulating such vital cellular processes as immune and inflammatory signaling and cancer.

Figure 5.37(A) depicts the proposed chemical mechanism for the HOIP-catalyzed formation of a linear poly-ubiquitin chain on the α-amino of Met-1 of a substrate protein.[55] It is obvious that the whole protein poly-ubiquitination cascade is fairly complex, involving not just the HOIP-catalyzed ligation reaction, but also the reactions respectively catalyzed by the E_1 ubiquitin-activating enzyme and the E_2 ubiquitin-conjugating enzyme, which serve to prepare for the HOIP-catalyzed reaction. Interestingly, different cysteine residues play an essential catalytic role in the catalysis by all the three types of enzymatic reactions. Given this, a fragment-based covalent ligand screening campaign was implemented from which the covalent inhibitor depicted in Figure 5.37(B) was identified as a potent, cell permeable, and likely proteome-wide selective irreversible inhibitor for the HOIP-catalyzed reaction.[55] The side chain Michael addition-type *S*-alkylation of a HOIP active site catalytic cysteine residue by the depicted covalent inhibitor could be that responsible for the experimentally observed irreversible inhibition of the HOIP-catalyzed reaction by the inhibitory compound.

References

1. G. A. Kotzia and N. E. Labrou, *Eur. J. Biochem.*, 2004, **271**, 3503.
2. Y. Q. Shi, S. Furuyoshi, I. Hubacek and R. R. Rando, *Biochemistry*, 1993, **32**, 3077.
3. J. Wang, S. Bauman and R. F. Colman, *Biochemistry*, 1998, **37**, 15671.
4. R. L. Edelstein and M. D. Distefano, *Biochem. Biophys. Res. Commun.*, 1997, **235**, 377.
5. V. Chowdhry, R. Vaughan and F. H. Westheimer, *Proc. Natl. Acad. Sci. U. S. A.*, 1976, **73**, 1406.
6. M. Worth, C.-W. Hu, H. Li, D. Fan, A. Estevez, D. Zhu, A. Wang and J. Jiang, *Chem. Commun.*, 2019, **55**, 13291.
7. T. M. Gloster, W. F. Zandberg, J. E. Heinonen, D. L. Shen, L. Deng and D. J. Vocadlo, *Nat. Chem. Biol.*, 2011, **7**, 174.
8. Y. Shishido, F. Tomoike, K. Kuwata, H. Fujikawa, Y. Sekido, Y. Murakami-Tonami, T. Kameda, N. Abe, Y. Kimura, S. Shuto and H. Abe, *ChemBioChem*, 2019, **20**, 900.
9. N. E. Labrou, D. J. Rigden and Y. D. Clonis, *Eur. J. Biochem.*, 2000, **267**, 6657.
10. A. M. Cesura, J. Gottowik, H. W. Lahm, G. Lang, R. Imhof, P. Malherbe, U. Röthlisberger and M. Da Prada, *Eur. J. Biochem.*, 1996, **236**, 996.
11. P. F. Kador, Y. S. Lee, L. Rodriguez, S. Sato, A. Bartoszko-Malik, Y. S. Abdel-Ghany and D. D. Miller, *Bioorg. Med. Chem.*, 1995, **3**, 1313.

12. P. Lee and R. F. Colman, *Bioconjugate Chem.*, 2005, **16**, 650.
13. R. Rajasekharan, R. C. Marians, J. M. Shockey and J. D. Kemp, *Biochemistry*, 1993, **32**, 12386.
14. M. F. Taylor, A. K. Bhattacharyya, K. Rajagopalan, R. Hiipakka, S. Liao and D. C. Collins, *Steroids*, 1996, **61**, 323.
15. M. de Rosa, S. Nonnis and A. Aliverti, *Biochem. Biophys. Res. Commun.*, 2021, **577**, 89.
16. P. A. Karplus, M. J. Daniels and J. R. Herriott, *Science*, 1991, **251**, 60.
17. G. Cullia, S. Bruno, S. Parapini, M. Margiotta, L. Tamborini, A. Pinto, A. Galbiati, A. Mozzarelli, M. Persico, A. Paladino, C. Fattorusso, D. Taramelli and P. Conti, *ACS Med. Chem. Lett.*, 2019, **10**, 590.
18. S. Bruno, A. Pinto, G. Paredi, L. Tamborini, C. De Micheli, V. La Pietra, L. Marinelli, E. Novellino, P. Conti and A. Mozzarelli, *J. Med. Chem.*, 2014, **57**, 7465.
19. M. Montefiori, F. S. Jørgensen and L. Olsen, *ACS Omega*, 2017, **2**, 4237.
20. C. Coelho, A. Foti, T. Hartmann, T. Santos-Silva, S. Leimkühler and M. J. Romão, *Nat. Chem. Biol.*, 2015, **11**, 779.
21. J. F. Alfaro, C. A. Joswig-Jones, W. Ouyang, J. Nichols, G. J. Crouch and J. P. Jones, *Drug Metab. Dispos.*, 2009, **37**, 2393.
22. J. Hosogi, R. Ohashi, H. Maeda, K. Fujita, J. Ushiki, T. Kuwabara, Y. Yamamoto and T. Imamura, *Biopharm. Drug Dispos.*, 2018, **39**, 164.
23. P. W. Thomas, M. Cammarata, J. S. Brodbelt, A. F. Monzingo, R. F. Pratt and W. Fast, *Biochemistry*, 2019, **58**, 2834.
24. H. Kojima, A. Kozaki, M. Iwata, T. Ando and Y. Kitade, *Bioorg. Med. Chem.*, 2008, **16**, 6575.
25. C. S. Yuan and R. T. Borchardt, *J. Biol. Chem.*, 1995, **270**, 16140.
26. K. Touhara and G. D. Prestwich, *J. Biol. Chem.*, 1993, **268**, 19604.
27. C. Ma, Z. Xia, M. D. Sacco, Y. Hu, J. A. Townsend, X. Meng, J. Choza, H. Tan, J. Jang, M. V. Gongora, X. Zhang, F. Zhang, Y. Xiang, M. T. Marty, Y. Chen and J. Wang, *J. Am. Chem. Soc.*, 2021, **143**, 20697.
28. W. Ren, R. Pengelly, M. Farren-Dai, S. Shamsi Kazem Abadi, V. Oehler, O. Akintola, J. Draper, M. Meanwell, S. Chakladar, K. Świderek, V. Moliner, R. Britton, T. M. Gloster and A. J. Bennet, *Nat. Commun.*, 2018, **9**, 3243.
29. P. W. Hruz, C. Narasimhan and H. M. Miziorko, *Biochemistry*, 1992, **31**, 6842.
30. Z. Fu, J. A. Runquist, C. Montgomery, H. M. Miziorko and J.-J. P. Kim, *J. Biol. Chem.*, 2010, **285**, 26341.
31. T. T. Lee, C. Worby, Z. Q. Bao, J. E. Dixon and R. F. Colman, *Biochemistry*, 1998, **37**, 8481.
32. M. Tsai, J. Koo, P. Yip, R. F. Colman, M. L. Segall and P. L. Howell, *J. Mol. Biol.*, 2007, **370**, 541.
33. N. E. Labrou, *J. Protein Chem.*, 1999, **18**, 729.
34. N. E. Labrou and Y. D. Clonis, *Arch. Biochem. Biophys.*, 1999, **365**, 17.
35. A. Basu, S. Subramanian and C. SivaRaman, *Biochemistry*, 1982, **21**, 4434.
36. B. X. C. Kwai, A. J. Collins, M. J. Middleditch, J. Sperry, G. Bashiri and I. K. H. Leung, *RSC Med. Chem.*, 2021, **12**, 57.

37. A. M. Bello, E. Poduch, M. Fujihashi, M. Amani, Y. Li, I. Crandall, R. Hui, P. I. Lee, K. C. Kain, E. F. Pai and L. P. Kotra, *J. Med. Chem.*, 2007, **50**, 915.
38. T. C. Appleby, C. Kinsland, T. P. Begley and S. E. Ealick, *Proc. Natl. Acad. Sci. U. S. A.*, 2000, **97**, 2005.
39. J. T. Stivers, C. Abeygunawardana, A. S. Mildvan, G. Hajipour, C. P. Whitman and L. H. Chen, *Biochemistry*, 1996, **35**, 803.
40. J. L. Thomas, R. P. Myers and R. C. Strickler, *J. Steroid Biochem. Mol. Biol.*, 1991, **39**, 471.
41. H. Park and K. M. Merz Jr., *J. Am. Chem. Soc.*, 2003, **125**, 901.
42. I. Abe, Y. F. Zheng and G. D. Prestwich, *Biochemistry*, 1998, **37**, 5779.
43. C. Hombrouck, M. L. Capmau and N. Moreau, *Cell. Mol. Biol.*, 1999, **45**, 347.
44. P. M. Hawkey, *J. Antimicrob. Chemother.*, 2003, **51** (Suppl. 1), 29.
45. K. D. Corbett and J. M. Berger, *Annu. Rev. Biophys. Biomol. Struct.*, 2004, **33**, 95.
46. L. R. McLean, Y. Zhang, H. Li, Z. Li, U. Lukasczyk, Y.-M. Choi, Z. Han, J. Prisco, J. Fordham, J. T. Tsay, S. Reiling, R. J. Vaz and Y. Li, *Bioorg. Med. Chem. Lett.*, 2009, **19**, 6717.
47. T. Soares, D. Goodsell, R. Ferreira, A. J. Olson and J. M. Briggs, *J. Mol. Recognit.*, 2000, **13**, 146.
48. E. Rosengren, R. Bucala, P. Aman, L. Jacobsson, G. Odh, C. N. Metz and H. Rorsman, *Mol. Med.*, 1996, **2**, 143.
49. P. de Aguiar Amaral, D. Autheman, G. D. de Melo, N. Gouault, J.-F. Cupif, S. Goyard, P. Dutra, N. Coatnoan, A. Cosson, D. Monet, F. Saul, A. Haouz, P. Uriac, A. Blondel and P. Minoprio, *PLoS Negl. Trop. Dis.*, 2018, **12**, e0006853.
50. L. Liu, R. Zhu, J. Li, Y. Pei, S. Wang, P. Xu, M. Wang, Y. Wen, H. Zhang, D. Du, H. Ding, H. Jiang, K. Chen, B. Zhou, L. Yu and C. Luo, *J. Med. Chem.*, 2022, **65**, 2174.
51. P. Rainey, E. Holler and M. R. Kula, *Eur. J. Biochem.*, 1976, **63**, 419.
52. A. C. Carr, G. L. Igloi, G. R. Penzer and J. A. Plumbridge, *Eur. J. Biochem.*, 1975, **54**, 169.
53. G. D. Wright and C. T. Walsh, *Protein Sci.*, 1993, **2**, 1765.
54. G. D. Wright and C. T. Walsh, *Acc. Chem. Res.*, 1992, **25**, 468.
55. H. Johansson, Y.-C. I. Tsai, K. Fantom, C.-W. Chung, S. Kümper, L. Martino, D. A. Thomas, H. C. Eberl, M. Muelbaier, D. House and K. Rittinger, *J. Am. Chem. Soc.*, 2019, **141**, 2703.

CHAPTER 6

Proteolysis Targeting Chimera (PROTAC) Design

6.1 Mode of Working

As described in other chapters in this book, different types of active site-directed covalent or non-covalent inhibitors have been successfully developed for various types of enzymatic reactions. However, it has been observed that the use of such inhibitors has been accompanied by the occurrence of (i) compensatory up-regulation of target enzyme abundance, which can be caused by a mechanism such as attenuated protein degradation[1–3] and/or (ii) compound resistance, which can be caused by the appearance of mutation(s) for the key amino acid residues in the target enzyme.[4,5] All these would compromise the pharmacological efficacy of the inhibitors. In view of these, a special type of the derivatives of the existing inhibitors, *i.e.* proteolysis targeting chimeras (PROTACs), were developed, aiming to eliminate the target enzymes *via* the ubiquitin-proteasome protein degradation system.[1,5–7]

As schematically illustrated in Figure 6.1, within the scope of this chapter a PROTAC molecule is a bivalent compound derived respectively from an active site-directed inhibitor (covalent or non-covalent) for a target enzyme to be degraded and a ligand able to bind and recruit an active endogenous E3 ubiquitin ligase to the proximity of the target enzyme, leading to degradation of the target enzyme. When designing a PROTAC molecule for an enzyme, the following must be considered: (i) the physicochemical properties of the linker need to be optimized to minimize the possible interference of the linker choice to the binding of the enzyme inhibitor and the E3 ubiquitin ligase ligand; and (ii) while the first E3 ubiquitin ligase ligands employed in constructing PROTAC molecules were peptide-based, small molecule-based ligands are being preferred, due to a foreseeable superior pharmacokinetic profile (*e.g.* enhanced cell permeability) of the corresponding PROTAC molecules as potential therapeutics and chemical biological research tools.

Active Site-directed Enzyme Inhibitors: Design Concepts
By Weiping Zheng

Published by the Royal Society of Chemistry, www.rsc.org

In order for an E3 ubiquitin ligase ligand to be a successful part in a PROTAC molecule, it is required to be able to bind to the E3 enzyme for example at the substrate binding site, however, this binding event should not inhibit the enzymatic activity of the E3 enzyme, in other words, the E3–ligand complex should be still catalytically active so as to bind and catalyze the poly-ubiquitination of the target enzyme in cooperation with the E2 ubiquitin-conjugating enzyme and the E1 ubiquitin-activating enzyme of the ubiquitin-proteasome protein degradation system.[8,9]

Even though there are ~600 E3 ubiquitin ligases in human cells, only a couple of them have been successfully and widely exploited in PROTAC design, *i.e.* those harboring substrate adaptors cereblon (CRBN) and von Hippel–Lindau (VHL). Small molecule-based ligands for CRBN (*e.g.* the immunomodulatory drug thalidomide and its structural derivatives lenalidomide and pomalidomide, which have all been approved by the United States Food and Drug Administration (FDA) for treating multiple myeloma) and VHL (*e.g.* VH-032) have been successfully and widely employed in PROTAC design.[1,8,10,11]

Unlike the stoichiometric binding by an active site-directed enzyme inhibitor, a PROTAC molecule would be able to sub-stoichiometrically or catalytically degrade the target enzyme, thereby able to exhibit a greater functional inhibition inside cells than the active site-directed enzyme inhibitor.[1,5,7] However, it has been observed that a PROTAC could be a stronger degrader of a target protein at a lower concentration than at a higher concentration; for example, the observed stronger SIRT2 degradation at 5 μM of PROTAC-A (depicted in Figure 6.4 in Section 6.2.1.1) in Michigan Cancer Foundation (MCF)-7 human breast cancer cells than that at 25 μM.[12] This is known as the "hook effect" with the PROTAC approach, which could result from the dis-favored bivalent binding *versus* the monovalent binding at a higher PROTAC concentration, *i.e.* the dis-favored formation of the functionally active ternary complex of a PROTAC with the target enzyme and E3 ubiquitin ligase necessary for the PROTAC-induced target protein poly-ubiquitination and proteasomal degradation *versus* the formation of the functionally inactive binary complex of the PROTAC with the target enzyme or E3 ubiquitin ligase.[13]

Another inviting feature of using a PROTAC molecule is the possibility of simultaneously deleting all the functions (*e.g.* various catalytic activities) associated with a target enzyme. Moreover, selective degradation among homologous enzymes catalyzing the same type of reaction would be also potentially achievable with the bivalent PROTAC molecules with different linker structures, which would be rooted in the different binding efficiencies with enzyme active sites and the E3 ubiquitin ligase.[14] In addition, cell type-selective degradation of a target enzyme could also be realized exploiting the cell type-selective expression of the target enzyme, *e.g.* selectively over-expressed in tumor *versus* non-tumor cells. Furthermore, due to the presence of two moieties in a PROTAC molecule responsible for binding respectively to the target enzyme and the E3 ubiquitin ligase, the use of a PROTAC molecule could lead to a potent and persistent degradation of a target

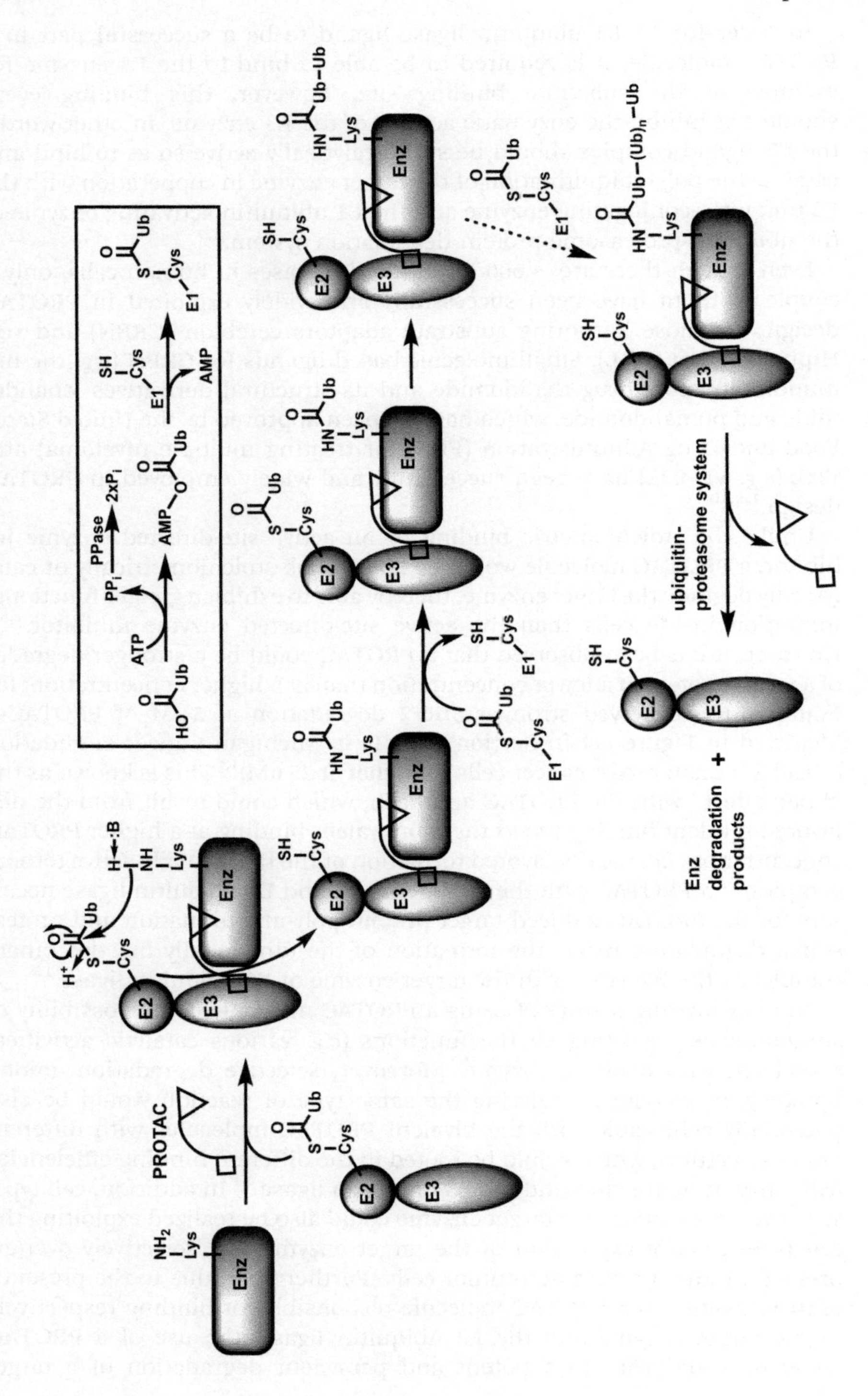
PROTAC
Enz
Lys
E2
E3
Cys
Ub
ATP
PPase
AMP
E1
SH
ubiquitin-
proteasome system
Enz
degradation
products

enzyme[14] or a simultaneous active site-directed inhibition of the target enzyme in case of its incomplete degradation, yet with a predictable net effect being prevention of the compensatory up-regulation of target enzyme abundance. However, because of this, the use of a PROTAC molecule could be more likely to lead to the occurrence of various side effects due to the possible functionally pleiotropic nature of a target enzyme.

As mentioned above, and elaborated below, small molecule-based ligands for CRBN (*e.g.* thalidomide, lenalidomide, and pomalidomide) and VHL (*e.g.* VH-032) have been those primarily employed in PROTAC design, however, the PROTAC molecules thus constructed still present unique problems to be overcome. Specifically, (i) the intracellular degradation of a target protein by such a PROTAC molecule may go faster or slower than the desirable intracellular protein turnover; (ii) the molecular weight and polar surface area of such a PROTAC molecule are typically greater than the common upper limits for a typical drug, especially an orally administered drug, which would justify the need for decreasing the values of these parameters of a PROTAC molecule so as to enhance its pharmacokinetic profile, *e.g.* cellular permeability and hydrophobicity/hydrophilicity balance; (iii) the expression level and enzymatic activity of the E3 ubiquitin ligases harboring CRBN or VHL are commonly low in organs such as kidney, lung, and brain, therefore, the PROTAC strategy reliant

Figure 6.1 A schematic illustration of the PROTAC-induced poly-ubiquitination and proteasomal degradation of an enzyme. Enz, enzyme; E3, ubiquitin ligase; E2, ubiquitin-conjugating enzyme; Lys, lysine; Cys, cysteine; Ub, ubiquitin (a 76-amino acid protein); ▽, an active site-directed enzyme inhibitor (covalent or non-covalent); □, a binder to the E3 ubiquitin ligase substrate binding site. Note: (i) the linker (⌒) between ▽ and □ has to be optimized to minimize the interference of the linker to the binding of ▽ and □, respectively, to the enzyme active site and the E3 ubiquitin ligase; (ii) while the binding of □ to E3 ubiquitin ligase substrate binding site would inhibit the binding, poly-ubiquitination, and proteasomal degradation of the substrate, this binding would not prevent the binding, poly-ubiquitination, and proteasomal degradation of the enzyme to be degraded; (iii) while the first □ employed in PROTAC molecules were peptide-based, small molecule-based □ are being preferred with a foreseeable superior pharmacokinetic profile of the corresponding PROTAC molecules; (iv) the number (n) of Ub units in a linear poly-ubiquitin chain is commonly two; and the inter-Ub linkage is an iso-peptide bond between the α-carboxyl of one Ub unit and the Lys side chain ε-amino of another Ub unit; (v) the Ub thioacylated E2 is formed from the E2-catalyzed thioacylation of a Cys side chain thiol (SH) on E2 with Ub thioacylated E1, which is formed *via* the E1 (ubiquitin-activating enzyme)-catalyzed activation of the Ub α-carboxyl followed by the thioacylation of a Cys side chain SH on E1 (inset); (vi) the E1-catalyzed reaction shown (inset) would be driven by the pyrophosphatase (PPase)-catalyzed hydrolysis of pyrophosphate (PP_i) to inorganic phosphate (P_i); (vii) this scheme is for the RING (really interesting new gene) family of E3 ligases (*e.g.* that harboring cereblon (CRBN) as the substrate adaptor) able to catalyze a direct transfer of Ub unit from the Ub thioacylated E_2 on to a substrate.

upon CRBN or VHL may not work well in the cells of these organs;[15–17] and (iv) extensive structure–activity optimization of the linker part of a lead PROTAC molecule thus constructed is often necessary.

Therefore, in order to overcome one or more of these problems, the following two derivative versions of the CRBN-/VHL-mediated PROTAC technology were introduced in recent years.[18,19] It should be noted that they have not yet been widely employed since their introduction into the field, however, they still deserve coverage here. It should be also noted that the example proteins used when developing these derivative approaches were not enzymes, *i.e.* bromodomain-containing protein 4 (Brd4) and estrogen-related receptor α (ERRα), but they fit well within this context for an illustrative purpose.

6.1.1 In-cell Construction of CRBN-mediated PROTACs for BRD4 and Extracellular Signal-regulated Kinase 1/2 (ERK1/2)

The right-hand half of Figure 6.2 shows how a PROTAC molecule could be generated inside cells from the two presumably cell permeable precursor molecules with the following example protein: BRD4, a protein able to specifically recognize the post-translational N^{ε}-acetyl-lysine and important in epigenetic regulation, *etc.* The left-hand half of Figure 6.2 shows the extended application of the approach to extracellular signal-regulated kinase 1/2 (ERK1/2), two closely related protein serine/threonine kinases engaged in the intracellular RAS protein signaling cascade implicated in multiple types of cancer.

Specifically, the depicted BRD4 inhibitor analog which bears a *trans*-cyclo-octene tag (circled portion) and the depicted thalidomide analog bearing the tetrazine tag (circled portion) were synthesized initially, and were then sequentially added to an incubate with a mammalian cell line, hoping the two presumably cell permeable precursor molecules would be able to enter cells and react with each other inside cells *via* the depicted bio-orthogonal inverse electron demand Diels–Alder cycloaddition reaction. Indeed, the product of this intracellular organic reaction, *i.e.* BRD4-PROTAC depicted in Figure 6.2, was found to be formed, and its generation inside cells would presumably be responsible for the observed BRD4 protein degradation inside cells, since the BRD4-PROTAC molecule constructed outside cells was found not to be capable of inducing BRD4 protein degradation inside cells presumably due to its cell impermeability. This cell impermeability might have originated from the presence of the polar hydrazine-like functionality in the bicyclic ring of BRD4-PROTAC at the reaction site of *trans*-cyclo-octene with tetrazine, which would hamper the cell penetration of BRD4-PROTAC. However, it should be made clear that many other PROTAC molecules synthesized outside cells are nevertheless cell permeable. As in the original version of the PROTAC approach (see below), the depicted monovalent ligands in Figure 6.2, *i.e.* the tagged BRD4 inhibitor and the tagged thalidomide, were also found not to be able to induce the BRD4 degradation inside cells, in contrast to the bivalent BRD4-PROTAC generated inside cells *per* the derivative PROTAC approach.

With the establishment of the derivative PROTAC approach with BRD4, it was then put into use with another protein, *i.e.* ERK1/2. Accordingly, the ERK1/2 inhibitor analog depicted in Figure 6.2 that bears a *trans*-cyclo-octene tag was also initially synthesized and was then added to an incubate with a mammalian cell line before the addition of the thalidomide analog tagged with tetrazine. The bivalent ERK1/2-PROTAC molecule depicted in Figure 6.2 was also found to be formed inside cells, presumably from the depicted intracellular bio-orthogonal inverse electron demand Diels–Alder cycloaddition reaction between the presumably cell permeable tagged thalidomide and tagged ERK1/2 inhibitor following their entry into cells. Presumably, the intracellularly generated ERK1/2-PROTAC would be also responsible for the observed ERK1/2 protein degradation inside cells, since the ERK1/2-PROTAC molecule pre-formed outside cells was found not to be capable of inducing ERK1/2 protein degradation inside cells presumably also due to its cell impermeability. This cell impermeability might have also originated from the presence of the polar hydrazine-like functionality in the bicyclic ring of ERK1/2-PROTAC at the reaction site of *trans*-cyclo-octene with tetrazine, which would hamper the cell penetration of ERK1/2-PROTAC, as described above for BRD4-PROTAC. The depicted monovalent ligands in Figure 6.2, *i.e.* the tagged ERK1/2 inhibitor and the tagged thalidomide, were also found not to be able to induce ERK1/2 degradation inside cells, in contrast to the bivalent ERK1/2-PROTAC generated inside cells *per* the derivative PROTAC approach.

It should be noted that the ERK1/2 inhibitor and its tagged analog shown in Figure 6.2 interact with ERK1/2 covalently with their acrylamide-like functionality serving as the Michael addition site with a nucleophile (cysteine side chain thiol (SH)) from ERK1/2 active site.[20] As indicated in Figure 6.2, this mode of interaction could be reasonably deemed to be conservable from the ERK1/2 inhibitor to its tagged analog and to ERK1/2-PROTAC. Therefore, while the observed ability of ERK1/2-PROTAC to induce ERK1/2 degradation inside cells attests to the feasibility of employing an electrophilic irreversible inhibitor to construct a PROTAC molecule with the same electrophilic functionality for a target protein, such obtained PROTAC molecule might be less proficient in degrading the protein inside cells when compared to the intracellular protein degradation induced by a PROTAC molecule based on a reversible inhibitor of the target protein. This is because the latter class of PROTAC molecules would be able to catalytically induce intracellular protein degradation, whereas it would be impossible for the former class of the electrophilic PROTAC molecules to behave catalytically when degrading an intracellular protein.

6.1.2 N-end Rule-enabled Single Destabilizing Amino Acid-mediated PROTACs for ERRα

In this derivative version of the PROTAC technology, one of the 13 so-called destabilizing amino acid residues[21–23] was incorporated into a PROTAC

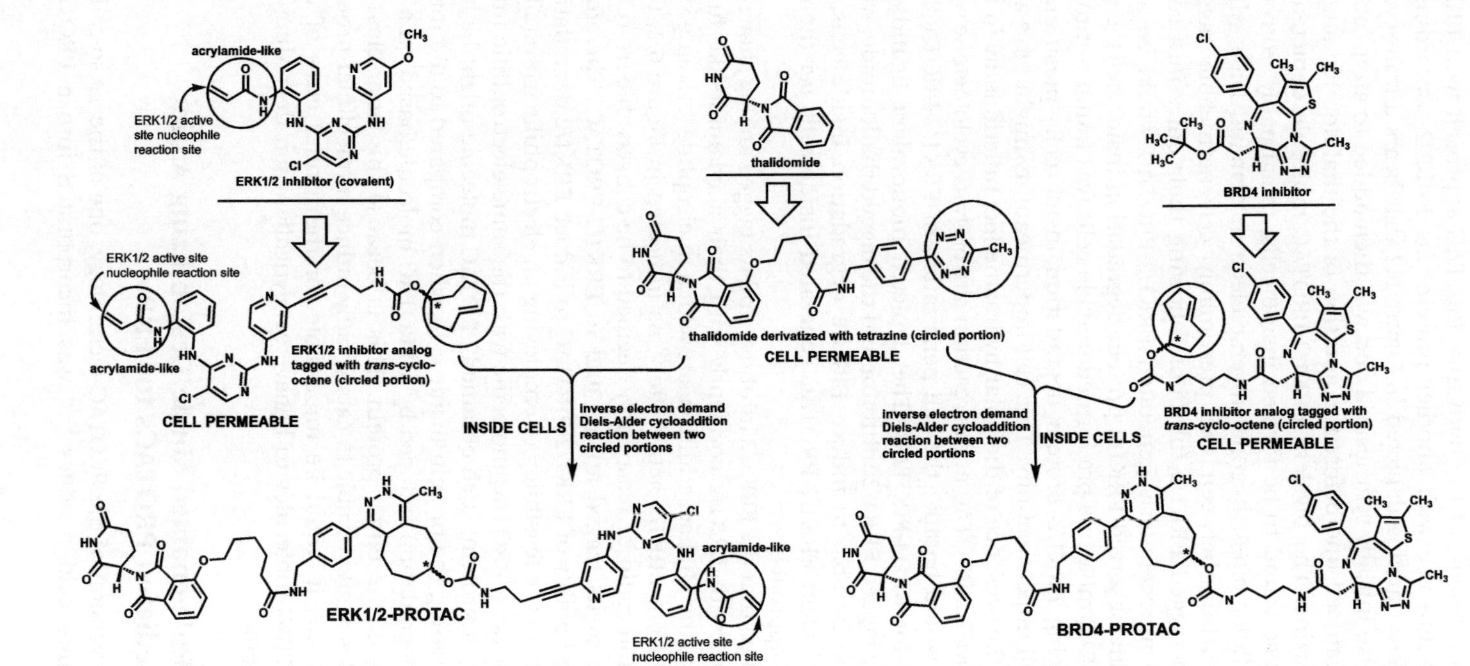

Figure 6.2 Right-hand half: The generation of BRD4-PROTAC inside cells from the tagged BRD4 inhibitor and the tagged thalidomide, two presumably cell permeable precursor molecules. Left-hand half: The generation of ERK1/2-PROTAC inside cells from the tagged thalidomide and the tagged ERK1/2 inhibitor which is another presumably cell permeable precursor molecule. Note: (i) the ERK1/2 inhibitor is a covalent inhibitor; (ii) the nucleophile supposed to interact covalently (Michael addition) and conservatively with the acrylamide-like functionality of the ERK1/2 inhibitor, its tagged analog, and ERK1/2-PROTAC is the side chain thiol (SH) of a cysteine residue at ERK1/2 active site; (iii) the four stereogenic carbons each indicated with an asterisk on the *trans*-cyclo-octene ring and the cyclo-octane ring are all racemic.

molecule, hoping that such an amino acid residue would be recognized by the really interesting new gene (RING) family E3 ubiquitin ligase Ubr1, leading to the target protein degradation inside cells *via* the N-end rule pathway, which is a ubiquitin-proteasome protein degradation system.[22,23] Of note, the N-end rule is named after the ability of an N-terminally exposed destabilizing amino acid residue to destabilize the protein *via* targeting it for proteasomal degradation.[22,23]

Figure 6.3 shows two such PROTAC molecules (*i.e.* ERRα-PROTAC-Arg and ERRα-PROTAC-His) harboring arginine (Arg) and histidine (His), respectively, two destabilizing amino acid residues. However, the depicted potent and selective ERRα inverse agonist *per in vitro* biochemical assays[24] was incorporated into both PROTAC molecules, serving as the binding ligand for ERRα. It can be seen from Figure 6.3 that this version of the PROTAC molecule employs a single destabilizing amino acid residue to recruit an E3 ubiquitin ligase (*i.e.* Ubr1) to realize the target protein degradation inside cells.

ERRα-PROTAC-Arg and ERRα-PROTAC-His were both found to efficiently degrade ERRα inside cells, however, the latter compound was found to be a more proficient ERRα degrader, suggesting that different destabilizing amino acid residues could exhibit different capability in inducing intracellular protein degradation. Moreover, ERRα-PROTAC-Arg and ERRα-PROTAC-His were both found to efficiently inhibit breast cancer cell growth.

6.2 Applications

As for the topic organization of Chapters 1–5, notable examples and the current status of development of the active site-directed inhibitor-derived PROTACs for each of the six types of enzymatic reactions (*i.e.* the reactions catalyzed by transferases, oxidoreductases, hydrolases, lyases, isomerases, or ligases) are elaborated herein. However, it should be noted that the target proteins to be degraded by the PROTAC technology can be those with or without catalytic activity, but this chapter is devoted only to those with catalytic activity (*i.e.* enzymes), and the inhibitors involved are only those that are active site-directed, with no discussion of other types of enzyme inhibitors such as allosteric inhibitors.

6.2.1 Applications with Transferases

In this section, notable examples of the active site-directed inhibitor-derived PROTACs for various transferase-catalyzed reactions are elaborated. It should be noted that the E3 ubiquitin ligases harboring CRBN or VHL are those primarily employed when developing the PROTACs for the transferase-catalyzed reactions.

6.2.1.1 Two CRBN-mediated Active Site-directed Inhibitor-derived PROTACs for the SIRT2-catalyzed Deacylation Reaction

As described in Section 1.2.1.1 and Figure 1.1, the sirtuin family of the NAD^+-dependent protein N^ε-acyl-lysine deacylases can be regarded as a

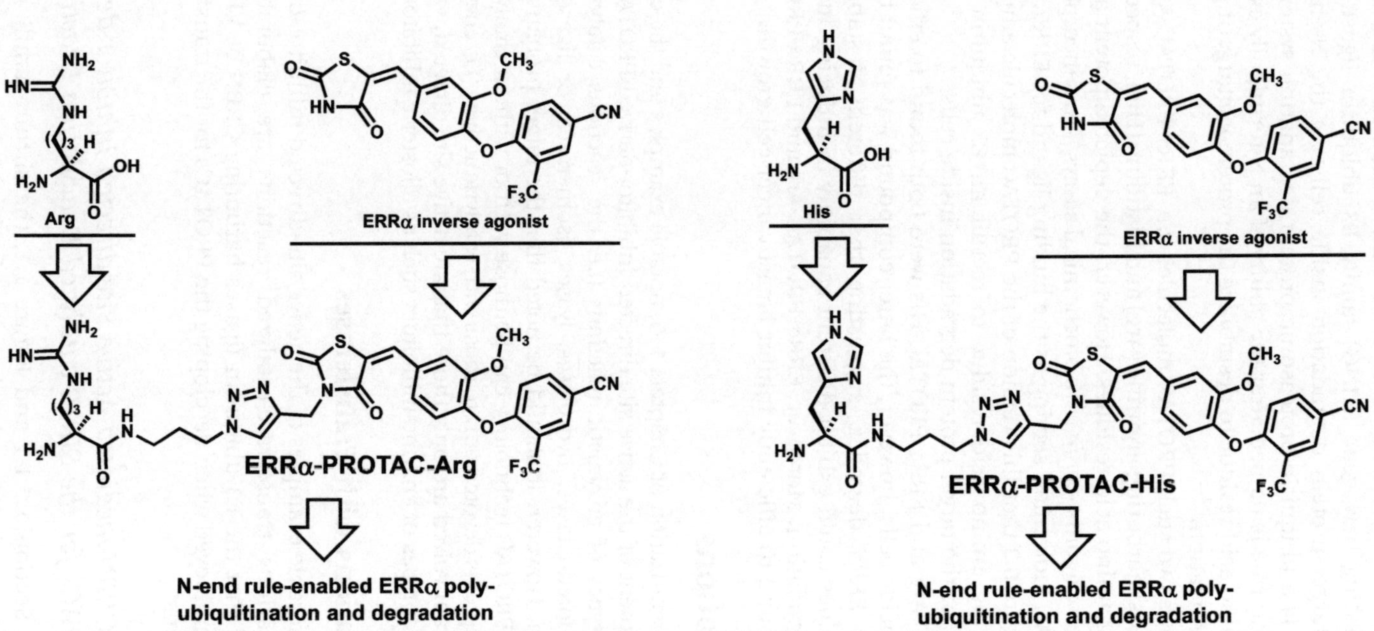

Arg

ERRα inverse agonist

ERRα-PROTAC-Arg

N-end rule-enabled ERRα poly-ubiquitination and degradation

His

ERRα inverse agonist

ERRα-PROTAC-His

N-end rule-enabled ERRα poly-ubiquitination and degradation

Figure 6.3 The chemical structures of ERRα-PROTAC-Arg and ERRα-PROTAC-His which are bivalent molecules each derived from the depicted ERRα inverse agonist (ERRα binding ligand) and a destabilizing amino acid residue (arginine (Arg) or histidine (His)) (binding ligands for the E3 ubiquitin ligase Ubr1). ERRα-PROTAC-Arg and ERRα-PROTAC-His both induce ERRα degradation inside cells *via* the N-end rule pathway, *i.e.* the Arg- or His-mediated recruitment of Ubr1 to poly-ubiquitinate ERRα for subsequent proteasomal degradation.

class of transferase enzymes since they catalyze an ADP-ribosylation reaction on the N^{ε}-acyl-lysine side chain amide oxygen at the beginning stage on the deacylation reaction coordinate and the net transfer of the acyl group from the N^{ε}-acyl-lysine substrate onto co-substrate NAD^+, with concomitant cleavage of the nicotinamide moiety from NAD^+ with the formation of the deacylated product and the acylated ADP-ribose (*i.e.* 2′-*O*-AADPR).

In mammals including humans, there are seven sirtuin isoforms, *i.e.* SIRT1–7, whose catalytic deacylation reactions play important roles in various life processes. As for SIRT2, it is primarily located in cytosol and is able to catalyze the removal of acetyl and fatty-acyl groups from the N^{ε}-acyl-lysine residue on substrate molecules. The SIRT2-catalyzed deacylation reaction is important in regulating vital cellular processes such as cell cycle progression and cellular metabolism, and holds a therapeutic potential for cancer and metabolic diseases. For a better functional deciphering and pharmacological exploitation of the SIRT2-catalyzed deacylation reaction, the following two CRBN-mediated PROTACs have been developed based upon the pre-existing active site-directed inhibitors for the SIRT2-catalyzed deacylation reaction.

Figure 6.4 shows the chemical structures of these two CRBN-mediated active site-directed inhibitor-derived PROTACs for the SIRT2-catalyzed deacylation reaction. PROTAC-A is a bivalent molecule composed of two covalently inter-linked structural motifs respectively derived from the small molecule-based CRBN ligand pomalidomide and the depicted active site-directed SIRT2 inhibitor TM (the compound code in the original literature),[25] which is a N^{ε}-thiomyristoyl-lysine-based catalytic mechanism-based potent and selective SIRT2 inhibitor whose mechanism of inhibition is presumably the SIRT2-catalyzed formation of the stalled catalytic intermediate III as depicted in Figure 1.2 following its uptake as an alternate substrate by the SIRT2 active site. PROTAC-B is a bivalent molecule composed of two covalently inter-linked structural motifs respectively derived from the small molecule-based CRBN ligand thalidomide and the depicted active site-directed SIRT2 inhibitor SirReal2 (the compound code in original literature)[26] which was discovered from a compound library screening campaign to be a potent and selective SIRT2 inhibitor. The mechanism of inhibition by SirReal2 appears to be the formation of a non-covalently stabilized tight binding rigid cyclic structure *via* an intra-molecular hydrogen bond between the amide N–H and one pyrimidine N in the structure of SirReal2 following its binding at the SIRT2 active site. To be consistent with the design rationale, unlike the monovalent compounds (*i.e.* CRBN ligand or SIRT2 inhibitor), PROTAC-A and PROTAC-B were both found to be able to efficiently degrade the SIRT2 protein and inhibit the SIRT2-catalyzed deacylation reaction inside cells. Moreover, PROTAC-A was further found to be able to potently inhibit both the deacetylase and the defatty-acylase activities of SIRT2, which would be unsurprising since the use of PROTAC-A would degrade the SIRT2 protein.

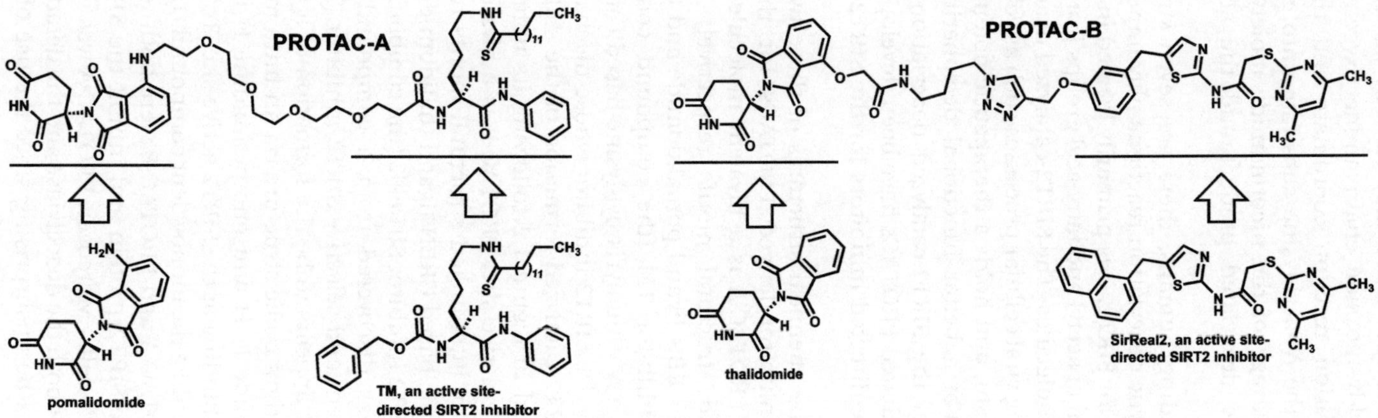

Figure 6.4 The chemical structures of PROTAC-A and PROTAC-B which are bivalent molecules each derived from the depicted active site-directed SIRT2 inhibitor and small molecule-based CRBN ligand (pomalidomide or thalidomide). TM and SirReal2 are the compound codes from the original literature reports.

6.2.1.2 A CRBN-mediated Active Site-directed Inhibitor-derived PROTAC for the BCR-ABL-catalyzed Tyrosine O*-Phosphorylation Reaction*

The breakpoint cluster region-Abelson (BCR-ABL) oncogenic fusion protein is an ATP-dependent protein tyrosine kinase whose enzymatic tyrosine *O*-phosphorylation reaction, as shown in Figure 6.5(A), plays an essential role for the initiation and/or maintenance of chronic myeloid leukemia (CML).[27] Therefore, the BCR-ABL-catalyzed phosphorylation reaction has been regarded as a valuable therapeutic target for CML. For a better pharmacological exploitation of the BCR-ABL-catalyzed phosphorylation reaction, the CRBN-mediated PROTAC compound SIAIS056 (the compound code in the original literature) depicted in Figure 6.5(B) has been developed based upon the pre-existing active site-directed potent inhibitor dasatinib[28] for this enzymatic reaction.[27] As shown, SIAIS056 is a bivalent molecule composed of two covalently inter-linked structural motifs derived from the small molecule-based CRBN ligand thalidomide and the depicted active site-directed inhibitor dasatinib.

Unlike the monovalent compounds (*i.e.* CRBN ligand or BCR-ABL inhibitor), SIAIS056 was found to efficiently degrade the BCR-ABL protein and thus inhibit the BCR-ABL-catalyzed phosphorylation reaction inside cells. Moreover, SIAIS056 treatment was found to produce a significant regression of a tumor xenograft of the CML patient-derived K562 cells. SIAIS056 was further found to also degrade certain clinically relevant dasatinib-resistant BCR-ABL mutants that still maintain a favorable binding interaction with dasatinib.

6.2.1.3 A CRBN-mediated Active Site-directed Inhibitor-derived PROTAC for the Bruton's Tyrosine Kinase-catalyzed Tyrosine O*-Phosphorylation Reaction*

Like the BCR-ABL oncogenic fusion protein described in Section 6.2.1.2, Bruton's tyrosine kinase (BTK) is also an ATP-dependent protein tyrosine kinase and the scheme shown in Figure 6.5(A) can also be used to depict the BTK-catalyzed tyrosine *O*-phosphorylation reaction. The BTK-catalyzed phosphorylation reaction constitutes a therapeutic target for B-cell malignancies such as chronic lymphocytic leukemia (CLL) and whose irreversible active site-directed covalent inhibition by the FDA-approved ibrutinib (*via* covalent bond formation between the C481 side chain thiol (SH) at the ATP binding pocket of BTK and the acrylamide-like moiety of ibrutinib) has been a major treatment option for CLL.[29,30] However, a prevalent ibrutinib-resistant BTK mutant (*i.e.* BTK[C481S]), isolated in $>80\%$ of CLL patients, would be unable to behave like the wild-type BTK in the formation of a covalent bond with ibrutinib, since the mutant has a serine instead of a cysteine residue at position 481, thus leading to the loss of ibrutinib's potent irreversible inhibition exhibited against the wild-type BTK-catalyzed phosphorylation reaction.[31,32]

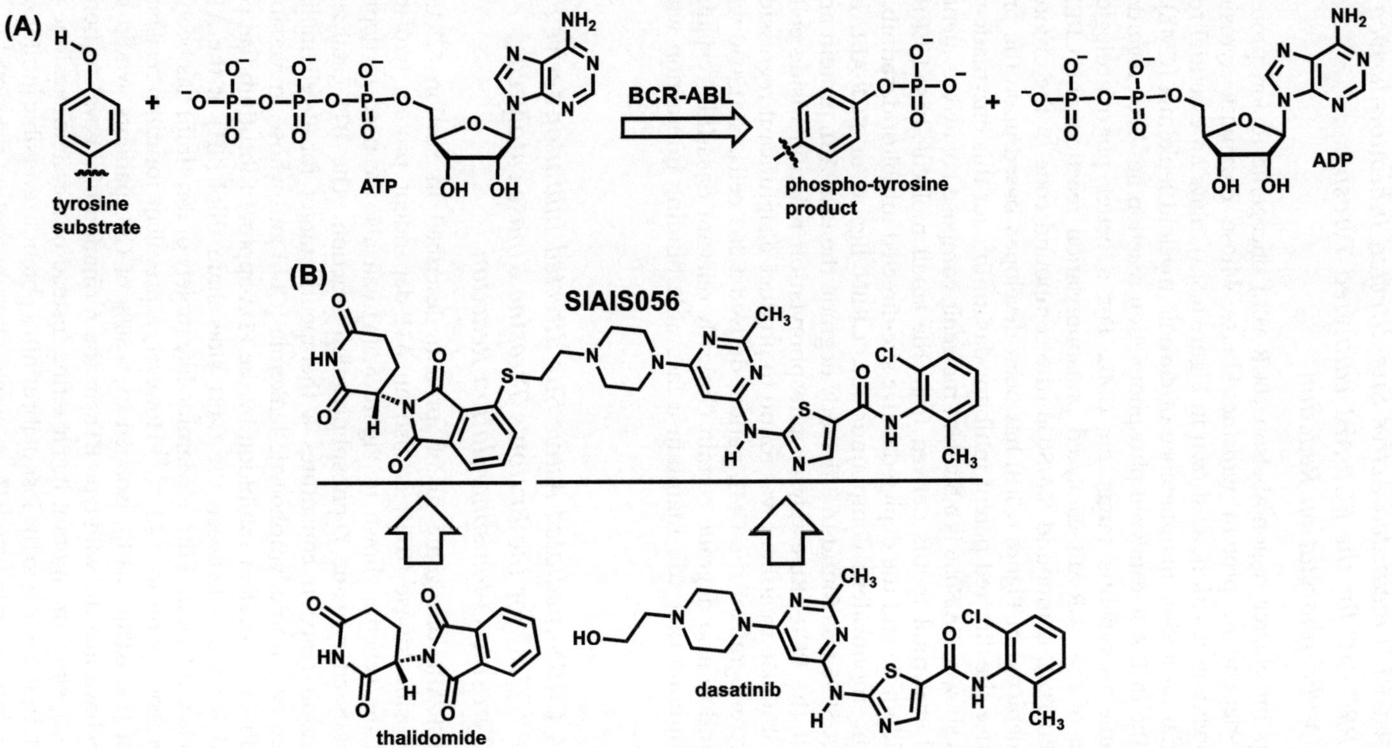

Figure 6.5 (A) The BCR-ABL-catalyzed ATP-dependent phosphotransfer reaction from the γ-phosphoryl group of ATP onto the tyrosine side chain hydroxyl group on a substrate molecule, with the formation of the phospho-tyrosine product and ADP. This scheme is similar to that shown in Figure 3.4(A). (B) The chemical structure of the PROTAC molecule SIAIS056, which is a bivalent molecule derived from the depicted active site-directed BCR-ABL inhibitor dasatinib and the small molecule-based CRBN ligand thalidomide. SIAIS056 is the compound code from the original literature report.

To develop an effective treatment for CLL harboring the BTK[C481S] mutant, the CRBN-mediated PROTAC molecule MT-802 (compound code in original literature) shown in Figure 6.6 was designed,[4] in the hope that it would be able to efficiently degrade not only the wild-type BTK, but also the BTK[C481S] mutant. Since a PROTAC molecule is a bivalent ligand, its overall binding affinity to the two protein binding partners (*i.e.* the enzyme to be degraded and an E3 ubiquitin ligase) in the ternary PROTAC complex would be greater than that of the two monovalent ligands toward the two protein binding partners, owing to the reduced binding entropic penalty for the pre-organized bivalent PROTAC molecule in comparison to the two monovalent ligands, as long as the construction of the PROTAC molecule from the two monovalent ligands would not hurt the binding of the PROTAC molecule to the two protein binding partners.[14] Therefore, the ternary PROTAC complex with the two protein binding partners could still be formed effectively from the two monovalent ligands with a weak binding affinity toward the two protein binding partners. Within this context, if ibrutinib still possesses a reasonable binding affinity toward the BTK[C481S] mutant, the ternary PROTAC complex could be still effectively formed, thus leading to an effective degradation of the BTK[C481S] mutant. Indeed, it was found that the wild-type BTK and the BTK[C481S] mutant were potently degraded inside cells with comparable degradation efficiency and kinetics by

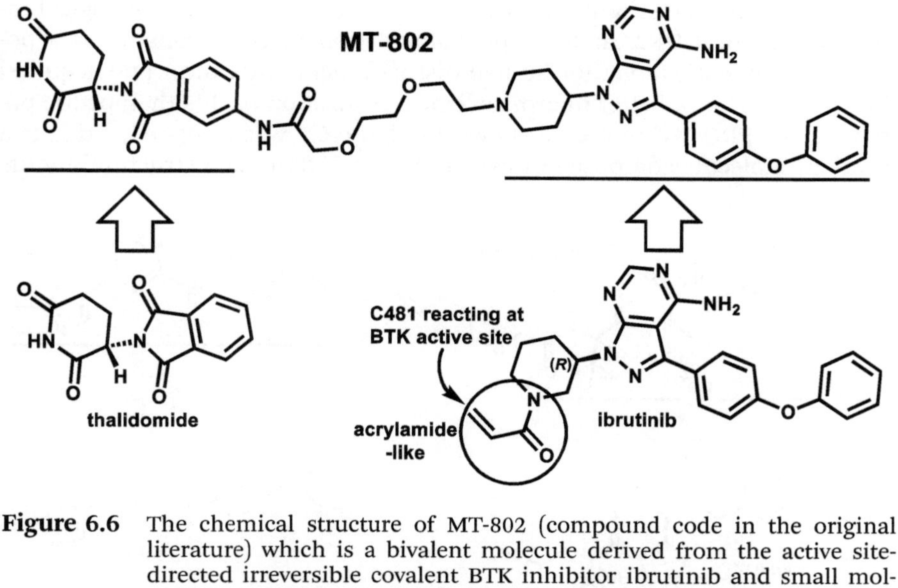

Figure 6.6 The chemical structure of MT-802 (compound code in the original literature) which is a bivalent molecule derived from the active site-directed irreversible covalent BTK inhibitor ibrutinib and small molecule-based CRBN ligand thalidomide. Note: (i) while ibrutinib is a covalent BTK inhibitor due to the presence of an acrylamide-like moiety in its structure able to covalently react with a BTK active site cysteine side chain thiol (SH), MT-802 does not have such an electrophilic moiety in its structure; (ii) the *R*-stereoconfiguration at the stereogenic carbon in the piperidine ring of ibrutinib is indicated.

the PROTAC molecule MT-802 whose treatment was further shown to inhibit the BTK signaling inside the B-cells derived from the CLL patients harboring the BTK[C481S] mutant.[4]

It is worth noting that treatment with up to 2.5 μM of MT-802 was not found to produce a significant "hook effect",[4] which could suggest that MT-802 is able to bring together BTK (wild-type or C481S mutant) and CRBN efficiently with a significant positive cooperativity promoted by a strong inter-protein interaction between BTK and CRBN, leading to the formation of a high-affinity functionally active PROTAC ternary complex and the minimized occurrence of the functionally inactive binary complexes with BTK or CRBN.

It should be also noted that, due to the lack of an electrophilic functionality (*e.g.* the acrylamide-like moiety in ibrutinib), MT-802 could be pharmacologically safer than ibrutinib in that it would have fewer off-targets and produce fewer adverse side-effects than the electrophilic ibrutinib.

6.2.1.4 *A VHL-mediated Active Site-directed Inhibitor-derived PROTAC for the Cyclin-dependent Kinase 2-Catalyzed Tyrosine* O*-Phosphorylation Reaction*

Cyclin-dependent kinase 2 (CDK2) is also an ATP-dependent protein tyrosine kinase whose enzymatic reaction (*i.e.* tyrosine *O*-phosphorylation) can be illustrated with the scheme shown in Figure 6.5(A). The pharmacological inhibition of the CDK2-catalyzed phosphorylation reaction constitutes a potential therapeutic target for human diseases including cancer and acquired hearing loss. For a better pharmacological realization of this therapeutic potential, the PROTAC molecule shown in Figure 6.7 was designed;[33] this is a bivalent molecule composed of two covalently inter-linked structural motifs

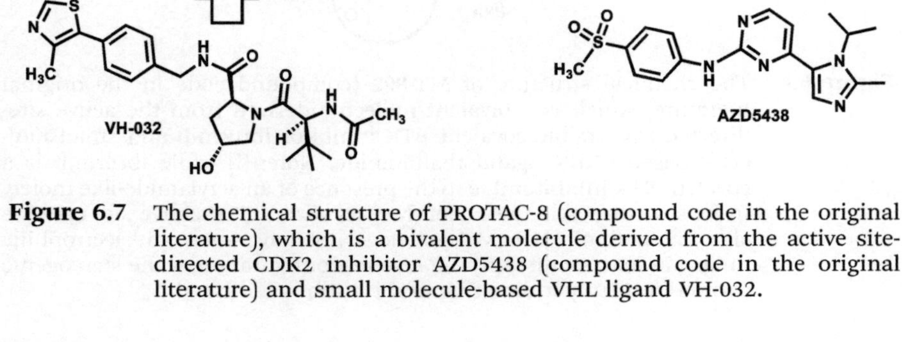

PROTAC-8

VH-032

AZD5438

Figure 6.7 The chemical structure of PROTAC-8 (compound code in the original literature), which is a bivalent molecule derived from the active site-directed CDK2 inhibitor AZD5438 (compound code in the original literature) and small molecule-based VHL ligand VH-032.

derived from the small molecule-based VHL ligand VH-032 and AZD5438 (the compound code in the original literature, an active site-directed, potent, yet non-selective inhibitor for the CDK2-catalyzed phosphorylation reaction[34]).

It was fairly gratifying to find that, while the monovalent AZD5438 was not able to significantly degrade CDK2, the bivalent PROTAC-8 behaved as a potent, and more importantly, selective CDK2 degrader (*versus* the closely related CDK1, CDK5, CDK7, and CDK9) inside cells.[33] This finding thus lends further support to the notion that a specific degrader for a given target enzyme can be effectively developed from a non-selective inhibitor for the same target enzyme among homologous enzymes, which could have been made so by the beauty of the bivalent ligand strategy[14] as for enzyme inhibition, that is selective inhibition for a given enzyme among homologous enzymes could be realized by simply adjusting how the two monovalent ligands are covalently connected so that different homologous enzymes have their own preferred relative positioning of the two monovalent motifs in a bivalent ligand. With *in vivo* zebrafish models of cisplatin-induced ototoxicity and kainic acid-induced excitotoxicity, like AZD5438, PROTAC-8 was found to also exert a protective role upon the neuromast hair cells against ototoxicity.

6.2.2 Applications with Oxidoreductases

In this section, notable examples of the active site-directed inhibitor-derived PROTACs for various oxidoreductase-catalyzed reactions are elaborated. It should be noted that the E3 ubiquitin ligases harboring CRBN or VHL have been employed when developing the PROTACs for oxidoreductase-catalyzed reactions.

6.2.2.1 A CRBN-mediated Active Site-directed Inhibitor-derived PROTAC for the Indoleamine 2,3-Dioxygenase 1-Catalyzed Reaction

As shown in Figure 6.8(A), indoleamine 2,3-dioxygenase 1 (IDO1) catalyzes the heme-dependent oxidative cleavage of the essential amino acid tryptophan to form *N*-formyl-kynurenine, which is the initial and rate-limiting step in the kynurenine pathway of tryptophan catabolism.[35,36] IDO1 is over-expressed in several cancers, such as colon cancer, melanoma, and ovarian cancer, and the IDO1-catalyzed reaction plays an important role in mediating the immune escape of cancer cells and has thus been regarded as an attractive target for cancer immunotherapy.

For a better pharmacological exploitation of the IDO1-catalyzed reaction, the bivalent PROTAC molecule (*i.e.* IDO1-PROTAC shown in Figure 6.8(B)) was designed based on the small molecule-based CRBN ligand thalidomide and the active site-directed potent IDO inhibitor epacadostat.[35,36] Of note, the coordination of epacadostat's *N*-hydroxyamidine hydroxyl (OH) to the heme center Fe^{2+} defines the key binding interaction of epacadostat with

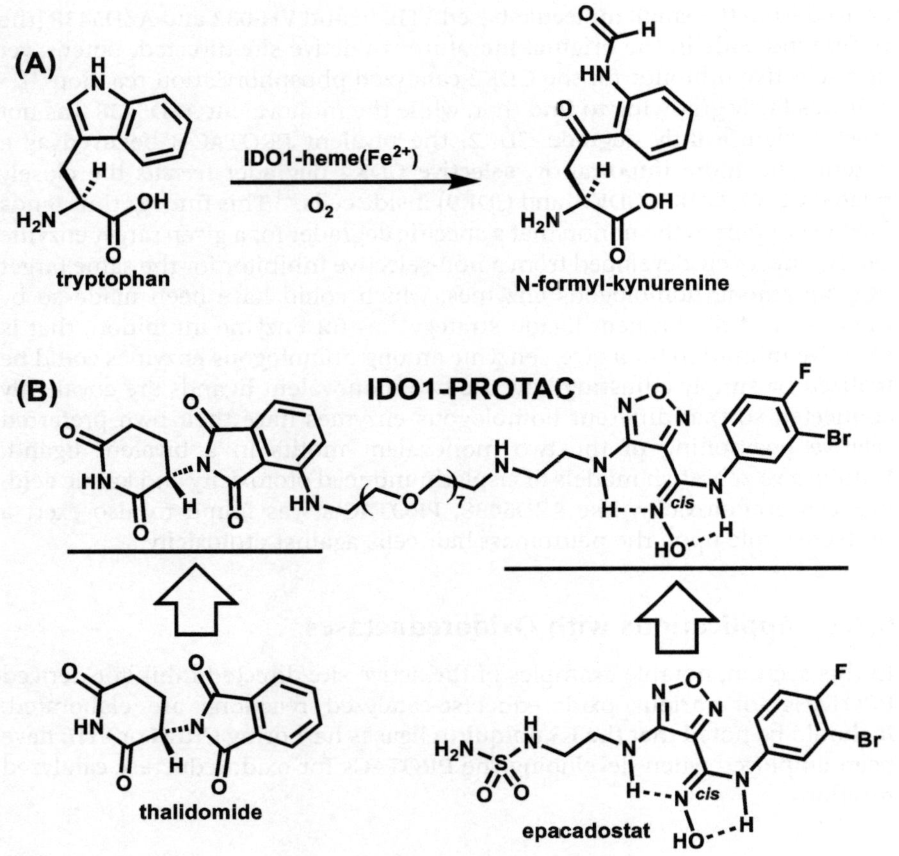

Figure 6.8 (A) A schematic illustration of the indoleamine 2,3-dioxygenase 1 (IDO1)-catalyzed heme-dependent oxidative cleavage of tryptophan to form *N*-formyl-kynurenine. (B) The chemical structure of IDO1-PROTAC, which is a bivalent molecule designed based on the small molecule-based CRBN ligand thalidomide and the active site-directed IDO inhibitor epacadostat. Hydrogen bonds are indicated by dashed lines.

IDO1. Unlike the monovalent ligands (*i.e.* IDO1 inhibitor and CRBN ligand), the bivalent IDO1-PROTAC was found to exert a potent and persistent degradation of the IDO1 protein inside cells.

6.2.2.2 A VHL-mediated Active Site-directed Inhibitor-derived PROTAC for the 3-Hydroxy-3-methylglutaryl Coenzyme A Reductase-catalyzed Reaction

3-Hydroxy-3-methylglutaryl coenzyme A (HMG-CoA) reductase (HMGCR) is an eight-transmembrane protein on the endoplasmic reticulum (ER) with its catalytic domain facing the cytosolic side of the ER membrane and catalyzing the initial and rate-limiting step in the mevalonate pathway of cholesterol biosynthesis, *i.e.* the NADPH-dependent conversion of the cytosolic HMG-CoA to mevalonate *via* the indicated catalytic intermediate, as shown in Figure 6.9(A).[3]

The inhibition of the HMGCR-catalyzed reaction has been a well-established therapeutic strategy for developing hypo-lipidemic drugs, and statins (*e.g.* lovastatin acid, shown in Figure 6.9(B)) have been identified as a major class of catalytic intermediate-based inhibitors for the HMGCR-catalyzed reaction, with the boxed part in an inhibitor molecule mimicking the boxed part in the catalytic intermediate shown in Figure 6.9(A). However, one drawback of the use of statins is the occurrence of compensatory up-regulation of HMGCR abundance,[3] which would compromise the pharmacological efficacy of the inhibitors. To circumvent this problem, the bivalent PROTAC molecule (*i.e.* HMGCR-PROTAC, shown in Figure 6.9(B)) was designed, which is composed of the covalently inter-linked VHL ligand VH-032 and the HMGCR inhibitor lovastatin acid.[3] HMGCR-PROTAC was found to be also a potent nano-molar inhibitor of the HMGCR-catalyzed reaction and to potently degrade HMGCR protein inside cells. A "hook effect" was also observed at higher concentrations of HMGCR-PROTAC, *i.e.* higher concentrations of HMGCR-PROTAC led to an enhanced HMGCR abundance.

Considering the charged nature of HMGCR-PROTAC, *i.e.* the negatively charged carboxylate under physiological pH, which would conceivably attenuate its cell permeability, the lactone pro-drug form of HMGCR-PROTAC (shown in Figure 6.9(C)) was developed and used in animal studies,[3] in the hope that the pro-drug would be converted to the active species HMGCR-PROTAC inside cells. It was found that oral administration of the lactone form of HMGCR-PROTAC with a diet-induced hypercholesterolemia mouse model not only potently degraded HMGCR protein, but also potently attenuated the cholesterol concentration.

6.2.2.3 *A CRBN-mediated Active Site-directed Inhibitor-derived PROTAC for the HMGCR-catalyzed Reaction*

As described in Section 6.2.2.2 and Figure 6.9(A), HMGCR catalyzes the NADPH-dependent transformation of the cytosolic HMG-CoA to mevalonate. The inhibition of this enzymatic reaction has been well established as a therapeutic strategy for developing hypo-lipidemic drugs, and statins have been identified as a major class of catalytic intermediate-based inhibitors for this enzymatic reaction.

As mentioned in Section 6.2.2.2, the administration of statins could lead to a compensatory up-regulation of HMGCR abundance. To circumvent this problem, the bivalent PROTAC molecule (*i.e.* HMGCR-PROTAC-2 shown in Figure 6.10) was also designed which is composed of the covalently inter-linked small molecule-based CRBN ligand thalidomide and the active site-directed potent HMGCR inhibitor atorvastatin.[2] HMGCR-PROTAC-2 was found to potently attenuate HMGCR protein concentration and potently

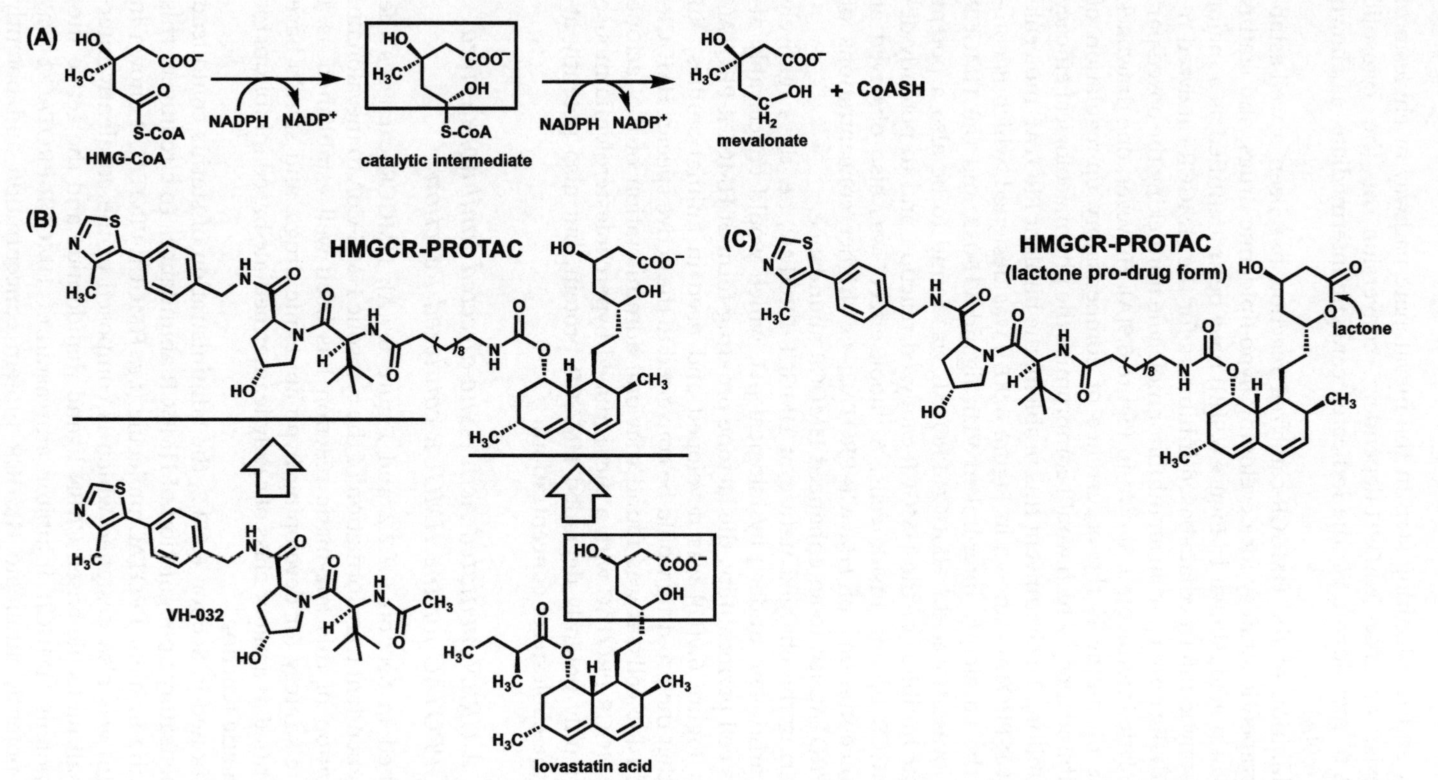

Figure 6.9 (A) A schematic illustration of the 3-hydroxy-3-methylglutaryl coenzyme A (HMG-CoA) reductase (HMGCR)-catalyzed NADPH-dependent reduction of HMG-CoA to mevalonate *via* the indicated catalytic intermediate (boxed structure). (B) The chemical structure of HMGCR-PROTAC, which is a bivalent molecule designed based on the small molecule-based VHL ligand VH-032 and the active site-directed catalytic intermediate-based HMGCR inhibitor lovastatin acid. (C) The chemical structure of the lactone pro-drug form of HMGCR-PROTAC.

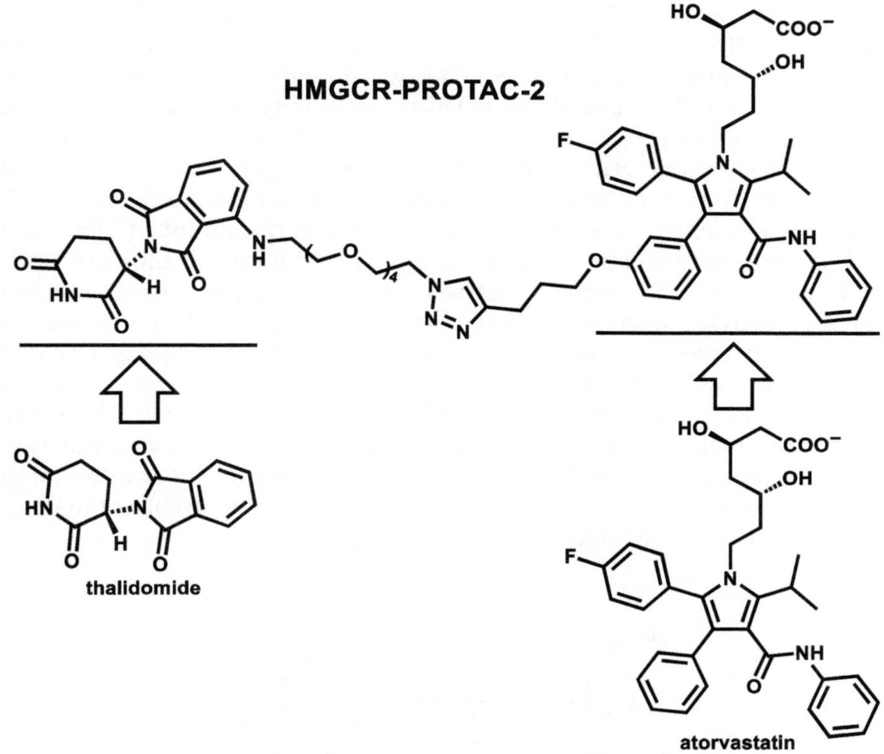

Figure 6.10 The chemical structure of HMGCR-PROTAC-2, which is a bivalent molecule derived from the small molecule-based CRBN ligand thalidomide and the active site-directed HMGCR inhibitor atorvastatin.

block cholesterol biosynthesis with diminished compensatory up-regulation of HMGCR abundance. Moreover, unlike the bivalent HMGCR-PROTAC-2, close monovalent structural analogs of atorvastatin were found to be incapable of inducing HMGCR degradation.

6.2.2.4 A VHL-mediated Active Site-directed Inhibitor-derived PROTAC for the Dihydroorotate Dehydrogenase-catalyzed Reaction

Dihydroorotate dehydrogenase (DHODH) is an enzyme located in the mitochondrial matrix and catalyzes the NAD^+-dependent oxidation of dihydroorotate to orotate, as depicted in Figure 6.11(A), which operates in the *de novo* biosynthetic pathway of uridine monophosphate.[37] The inhibition of the DHODH-catalyzed reaction has been regarded as a potential therapeutic strategy for cancers. For a better pharmacological exploitation of this enzymatic reaction, the bivalent PROTAC molecule (*i.e.* DHODH-PROTAC, shown in Figure 6.11(B)) was designed based on the small molecule-based VHL ligand VH-032 and the potent (nano-molar) active site-directed DHODH inhibitor brequinar.[37] DHODH-PROTAC was found to be a nano-molar inhibitor for the DHODH-catalyzed reaction. Moreover, in a cytotoxicity assay with the DHODH inhibition-sensitive colon cancer cell line HCT-116, the methyl ester form of DHODH-PROTAC (shown in Figure 6.11(B)) was found to be more than four-fold more cytotoxic than DHODH-PROTAC, which could have resulted from the higher cellular permeability of the more hydrophobic ester compound and the ability of the ester compound to be hydrolyzed into the active negatively charged carboxylate compound DHODH-PROTAC once inside cells. However, DHODH-PROTAC and its methyl ester form were both found to be incapable of inducing DHODH degradation inside cells, which could be due to the following: (i) DHODH is a mitochondrial matrix protein and such proteins are degraded within mitochondrial matrix by mechanism(s) quite different from those of the cytosolic proteasomal protein degradation system; and (ii) the linker/inhibitor of DHODH-PROTAC may need to be refined to make the DHODH-PROTAC-induced DHODH degradation become more possible.

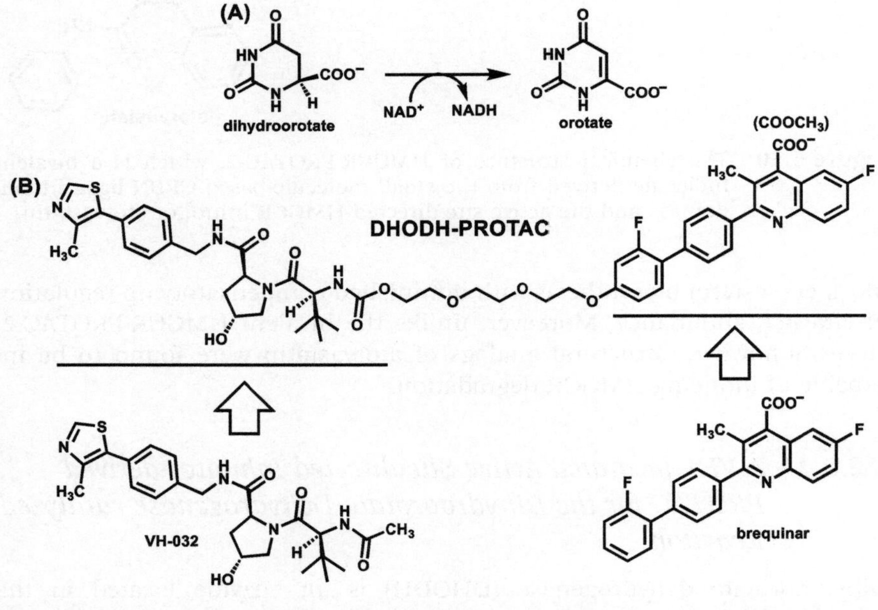

Figure 6.11 (A) A schematic illustration of the dihydroorotate dehydrogenase (DHODH)-catalyzed NAD^+-dependent oxidation of dihydroorotate to orotate. (B) The chemical structure of DHODH-PROTAC, which is a bivalent molecule designed based on the small molecule-based VHL ligand VH-032 and the active site-directed DHODH inhibitor brequinar. The methyl ester form of DHODH-PROTAC is also indicated.

6.2.3 Applications with Hydrolases: A CRBN-mediated Active Site-directed Inhibitor-derived PROTAC for the Histone Deacetylase 8-Catalyzed Reaction

In addition to the seven NAD^+-dependent sirtuin family of protein N^{ε}-acyl-lysine deacylases (*i.e.* SIRT1–7) described in Sections 1.2.1.1 and 6.2.1.1, mammals, including humans, also harbor 11 Zn^{2+}-dependent protein N^{ε}-acyl-lysine deacylases (*i.e.* histone deacetylase (HDAC) 1–11).[38,39] Of note, HDAC is an old yet still used term, and stands for histone deacetylase, since the first substrate proteins identified were histone proteins and the first type of N^{ε}-acyl-lysine deacylation identified was deacetylation.

HDAC8 is a HDAC enzyme able to catalyze the removal of acetyl and bulkier groups (*e.g.* myristoyl) from N^{ε}-acyl-lysine, as shown in Figure 6.12(A). It should be noted that HDAC8 may also use Fe^{2+} as the endogenous catalytic cofactor in addition to Zn^{2+}. HDAC8 is significantly over-expressed in certain cancer cells (*e.g.* A549 and G361 cells) and the HDAC8-catalyzed deacylation reaction plays an important role in regulating such vital life processes as cell

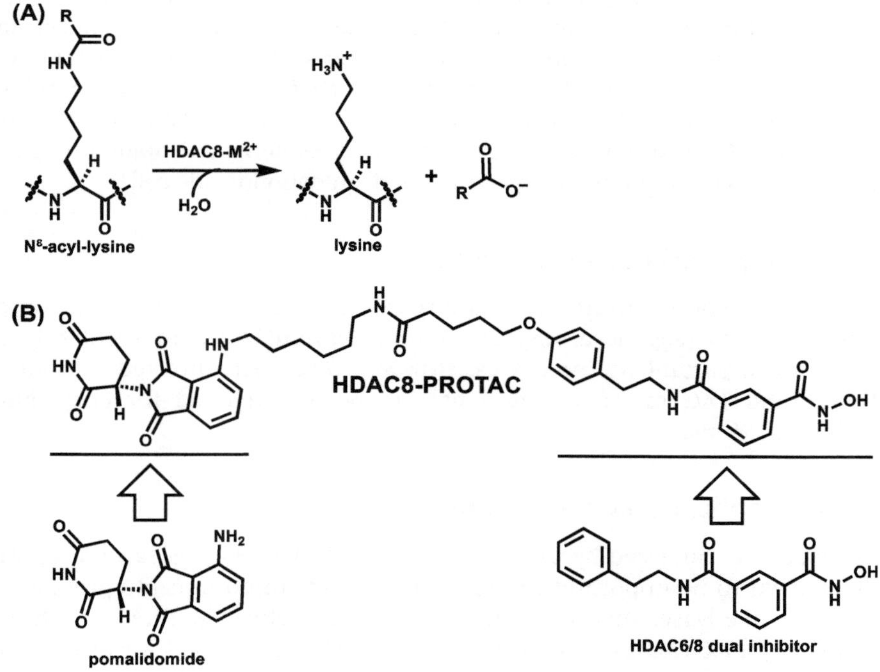

Figure 6.12 (A) A schematic illustration of the histone deacetylase 8 (HDAC8)-catalyzed M^{2+}-dependent N^{ε}-acyl-lysine deacylation reaction. R, acetyl or bulkier groups (*e.g.* myristoyl); M^{2+}, Zn^{2+} or Fe^{2+}. (B) The chemical structure of HDAC8-PROTAC, which is a bivalent molecule designed based on the small molecule-based CRBN ligand pomalidomide and the depicted active site-directed HDAC6/8 dual inhibitor.

cycle progression and tumor promotion. Therefore, inhibitors for the HDAC8-catalyzd deacylation reaction have been actively pursued as potential therapeutic agents for human diseases such as cancer.[38,39]

To facilitate the pharmacological exploitation of the HDAC8-catalzyed deacylation reaction, the bivalent PROTAC molecule (*i.e.* HDAC8-PROTAC shown in Figure 6.12(B)) was designed based on the small molecule-based CRBN ligand pomalidomide and a potent (nano-molar) active site-directed dual inhibitor for the enzymatic deacylation reactions respectively catalyzed by HDAC8 and HDAC6 (another mammalian HDAC enzyme).[38] It should be noted that, even though a HDAC6/8 dual inhibitor was used when designing HDAC8-PROTAC, a HDAC8-selective PROTAC molecule could still be obtained following a judicious manipulation of the physicochemical profile of the linker and its attaching site(s) on one or two monovalent moieties due to the bivalent binding nature of a PROTAC molecule relative to the two monovalent counterparts. Indeed, HDAC8-PROTAC was found not only to be a potent HDAC8 degrader inside cells, but also to be a selective HDAC8 degrader *versus* other HDAC enzymes including HDAC1/3/6. A "hook effect" was also observed, *i.e.* the HDAC8 degrading efficacy of HDAC8-PROTAC became attenuated at higher concentrations ($\geq$40 μM). Moreover, the monovalent ligands, *i.e.* the HDAC6/8 dual inhibitor and the CRBN ligand pomalidomide shown in Figure 6.12(B) were both found to be unable to degrade HDAC8 inside cells under the same experimental conditions under which HDAC8-PROTAC was found to potently degrade HADAC8, which would support the notion that an active ternary complex of a PROTAC molecule with a target enzyme and an E3 ubiquitin ligase has to be formed to promote target enzyme degradation inside cells.

6.2.4 Applications with Lyases

Enzymatic lyation (group elimination) reactions catalyzed by lyases and other enzymes (*e.g.* decarboxylase) play important roles in regulating a variety of important life processes. However, active site-directed inhibitor-derived PROTAC molecules have not yet been developed for enzymatic lyation reactions.

6.2.5 Applications with Isomerases

The reactions catalyzed by isomerases or other enzymes (*e.g.* racemase) are also known to be important in regulating various important life processes. However, like lyases mentioned in Section 6.2.4, active site-directed inhibitor-derived PROTAC molecules have not yet been developed for enzymatic isomerization reactions.

6.2.6 Applications with Ligases

In this section, notable examples of the active site-directed inhibitor-derived PROTACs for various ligase-catalyzed reactions are elaborated. It should be noted that the E3 ubiquitin ligase harboring CRBN has been the one primarily employed when developing PROTACs for ligase-catalyzed reactions.

6.2.6.1 *A CRBN-mediated Active Site-directed Inhibitor-derived PROTAC for the Murine Double Minute 2-Catalyzed Ligation Reaction*

Murine double minute 2 (MDM2) is an E3 ubiquitin ligase of the RING family able to catalyze the poly-ubiquitination of a protein substrate, signaling for its degradation by the ubiquitin-proteasome protein degradation system, as illustrated in Figure 6.13(A).[40] The major human tumor suppressor protein p53 (wild-type) is a substrate for the MDM2-catalyzed ligation reaction, and its enhanced degradation in MDM2 over-expressing tumor cells would promote tumorigenesis. Therefore, the inhibitors of the MDM2-catalyzed ligation reaction have been actively pursued as potential anti-cancer therapeutic agents and the wild-type p53 protein has been a

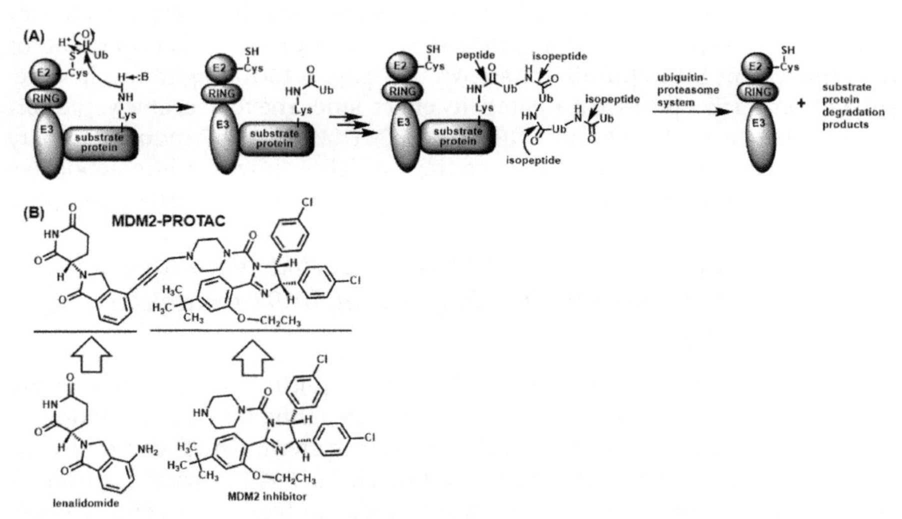

Figure 6.13 (A) A schematic illustration of the E3 ubiquitin ligase murine double minute 2 (MDM2)-catalyzed poly-ubiquitination of a protein substrate. E2, ubiquitin-conjugating enzyme; E3, ubiquitin ligase; RING, the really interesting new gene; :B, a general base at the MDM2 active site; Cys, cysteine; Lys, lysine; Ub, ubiquitin (a 76-amino acid protein). Note: (i) the number of Ub units in a linear poly-ubiquitin chain is commonly four, as shown; (ii) peptide bonds and isopeptide bonds in the linear poly-ubiquitin chain are both indicated, with isopeptide bonds being the inter-Ub linkage between the α-carboxyl of one Ub unit and the Lys side chain ε-amino of another Ub unit. (B) The chemical structure of MDM2-PROTAC, which is a bivalent molecule designed based on the small molecule-based CRBN ligand lenalidomide and the depicted active site-directed nutlin-type inhibitor for the MDM2-catalyzed ligation reaction.

valuable structural template for the inhibitor design. Indeed, a class of inhibitors known as nutlins have been identified as potent MDM2 inhibitors by virtue of their ability to assume an α-helical conformation mimicking that of the wild-type p53 protein substrate bound at the MDM2 active site.[40]

To facilitate the pharmacological exploitation of the MDM2-catalyzed ligation reaction, the bivalent PROTAC molecule (*i.e.* MDM2-PROTAC shown in Figure 6.13(B)) was designed based on the small molecule-based CRBN ligand lenalidomide and a potent active site-directed nutlin-type inhibitor for the MDM2-catalyzed ligation reaction.[40] MDM2-PROTAC was found to be capable of promoting an efficient degradation of MDM2 protein and a potent growth inhibition in RS4;11 leukemia cells. Moreover, MDM2-PROTAC was also found to be able to efficiently promote apoptosis in RS4;11 leukemia cells.

As described in Section 6.2.1.3 for MT-802, a strong positive cooperativity for the MT-802-promoted physical association between BTK and CRBN could have promoted the formation of a high-affinity functionally active PROTAC ternary complex and a minimized occurrence of the functionally inactive binary complexes with BTK or CRBN, leading to the insignificant occurrence of the "hook effect". As such, the much stronger MDM2 protein degrading efficacy observed for the bivalent MDM2-PROTAC with a very short linker over that for the monovalent MDM2 inhibitor shown in Figure 6.13(B) might also have resulted from such a positive cooperativity effect, since the foreseeable very short distance between MDM2 and CRBN in the MDM2-PROTAC-induced ternary complex would likely be able to encourage a tighter physical interaction between MDM2 and CRBN, thus more efficiently stabilizing the ternary complex.

6.2.6.2 A CRBN-mediated Active Site-directed Inhibitor-derived PROTAC for the Zinc Finger Protein 91-Catalyzed Ligation Reaction

Zinc finger protein 91 (ZFP91) is another E3 ubiquitin ligase able to catalyze the poly-ubiquitination of a protein substrate signaling for its degradation by the ubiquitin-proteasome protein degradation system, which could also be illustrated by the scheme in Figure 6.13(A). The ZFP91-catalyzed ligation reaction is oncogenic by virtue of its involvement in regulating cellular oncogenic signaling pathways such as those mediated by nuclear factor-κB (NF-κB) and hypoxia-inducible factor-1α (HIF-1α). Therefore, the inhibitors of the ZFP91-catalyzed ligation reaction would hold an anti-cancer therapeutic potential.

While there are no such reported inhibitors in the current literature, one CRBN-mediated PROTAC based on the signal transducer and transcription factor 3 (STAT3) inhibitor napabucasin, *i.e.* napabucasin-PROTAC shown in Figure 6.14, has been developed for the ZFP91-catalyzed ligation reaction.[41] It should be noted that, even though the possible inhibitory action of napabucasin against the ZFP91-catalyzed ligation reaction was not investigated in the literature, the observed strong degradation efficacy of napabucasin-PROTAC toward ZFP91 protein (see below) could support an educated guess

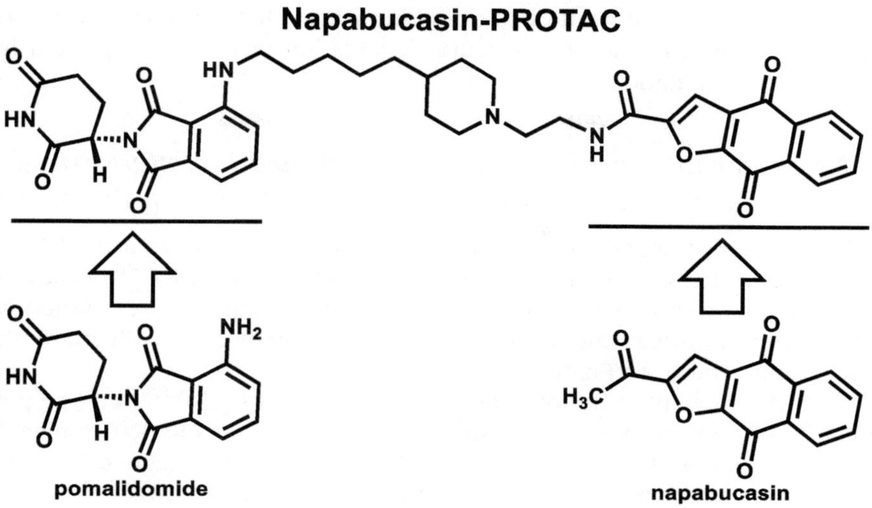

Napabucasin-PROTAC

pomalidomide

napabucasin

Figure 6.14 The chemical structure of napabucasin-PROTAC, which is a bivalent molecule designed based on the small molecule-based CRBN ligand pomalidomide and the signal transducer and transcription factor 3 (STAT3) inhibitor napabucasin, which is also likely an active site-directed zinc finger protein 91 (ZFP91) inhibitor.

that napabucasin could behave as an active site-directed inhibitor for the ZFP91-catalyzed ligation reaction.

Like napabucasin, napabucasin-PROTAC was found to be a STAT3 inhibitor, however, napabucasin-PROTAC was found to be incapable of degrading STAT3 inside cells, instead, this PROTAC molecule was able to potently degrade ZFP91 protein. It was further found that napabucasin-PROTAC was cytotoxic, however, the degradation of ZFP91 protein induced by this PROTAC molecule only partially attenuated its cytotoxicity, suggesting that its cytotoxicity is mediated by multiple cellular targets, not just ZFP91.

It is worth noting that the napabucasin-PROTAC-induced degradation of ZFP91 protein seems to also exhibit a "hook effect", *i.e.* a lower percentage of ZFP91 protein was degraded at higher concentrations of napabucasin-PROTAC, implying that there is a low positive cooperativity for the napabucasin-PROTAC-induced degradation of ZFP91 protein. Moreover, under the same experimental conditions, in contrast to the potent degradation of ZFP91 protein by the bivalent napabucasin-PROTAC, monovalent ligands (*i.e.* napabucasin and pomalidomide) were found to be much weaker degraders of ZFP91 protein. Another important observation also worth noting is that napabucasin-PROTAC was found not to degrade G1 to S phase transition protein (GSPT), suggesting that this PROTAC molecule is able to behave as a selective degrader toward ZFP91 protein. Since the binding of thalidomide and its derivatives lenalidomide and pomalidomide to CRBN can remodel the CRBN surface, making it be able to bind one or more so-called neo-substrates,[8,9,42] it would be an inviting feature for a proficient PROTAC molecule to be capable of a selective degradation of the target protein intended for degradation.

6.2.6.3 *A CRBN-mediated Active Site-directed Inhibitor-derived PROTAC for the CUL4^{CRBN}-catalyzed Ligation Reaction*

CRBN is a substrate adaptor for an E3 ubiquitin ligase known as CUL4^{CRBN} of the Cullin-RING sub-family of the RING family.[1] The CUL4^{CRBN}-catalyzed protein substrate poly-ubiquitination and the consequent degradation by the ubiquitin-proteasome protein degradation system can be illustrated by the scheme shown in Figure 6.13(A).

To further decipher the physiological functions of CUL4^{CRBN} and help to identify its further endogenous substrates, the bivalent PROTAC molecule (*i.e.* thalidomide-PROTAC) shown in Figure 6.15 was designed.[43] When interacting with this symmetrical PROTAC molecule, CUL4^{CRBN} simultaneously serves as the E3 ubiquitin ligase and the target protein to be degraded. Thalidomide-PROTAC was found to degrade CUL4^{CRBN} inside cells potently and selectively, since it only marginally degraded two CUL4^{CRBN} neo-substrates, *i.e.* the lymphoid transcription factors Ikaros family zinc fingers 1 and 3 (IKZF1 and IKZF3). Moreover, under the same experimental conditions, the monovalent ligand pomalidomide (a close derivative of thalidomide) was found to be unable to degrade CUL4^{CRBN}. It was further

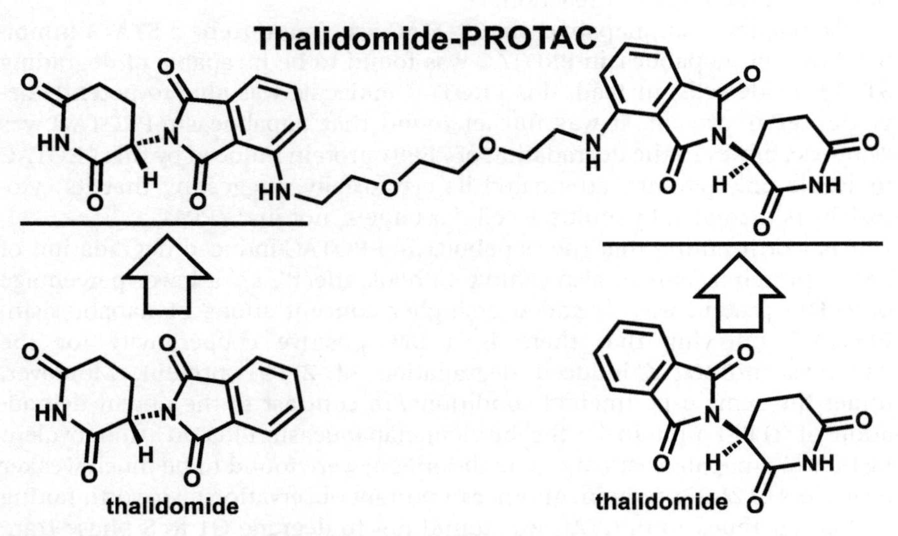

Figure 6.15 The chemical structure of thalidomide-PROTAC, which is a bivalent molecule designed based on the small molecule-based CRBN ligand thalidomide and thalidomide itself, which is also likely an active site-directed inhibitor for the E3 ubiquitin ligase CUL4^{CRBN}-catalyzed protein poly-ubiquitination reaction.

found that, while thalidomide-PROTAC treatment abrogated the cellular effects of thalidomide, lenalidomide, and pomalidomide in multiple myeloma cells, this treatment and the $CUL4^{CRBN}$ degradation did not exhibit an effect on the viability and proliferation of multiple myeloma cells. As such, thalidomide-PROTAC could be employed to identify further $CUL4^{CRBN}$ endogenous substrates and to further decipher the physiological functions of $CUL4^{CRBN}$, and to facilitate the cellular mechanistic investigation on the immunomodulatory drugs thalidomide, lenalidomide, and pomalidomide.

As mentioned above, CRBN is a substrate adaptor for $CUL4^{CRBN}$, therefore, thalidomide can be regarded as a binder to the substrate binding site of $CUL4^{CRBN}$. However, the possible inhibitory action of thalidomide against the $CUL4^{CRBN}$-catalyzed ligation reaction was not investigated in the literature. Nevertheless, when considering the observed high degradation efficacy of thalidomide-PROTAC toward the CRBN subunit of $CUL4^{CRBN}$, thalidomide could be regarded to behave as an active site-directed inhibitor for the $CUL4^{CRBN}$-catalyzed ligation reaction, and thalidomide-PROTAC would be an active site-directed inhibitor-derived PROTAC for the $CUL4^{CRBN}$-catalyzed ligation reaction.

References

1. P. Jevtić, D. L. Haakonsen and M. Rapé, *Cell Chem. Biol.*, 2021, **28**, 1000.
2. M.-X. Li, Y. Yang, Q. Zhao, Y. Wu, L. Song, H. Yang, M. He, H. Gao, B.-L. Song, J. Luo and Y. Rao, *J. Med. Chem.*, 2020, **63**, 4908.
3. G. Luo, Z. Li, X. Lin, X. Li, Y. Chen, K. Xi, M. Xiao, H. Wei, L. Zhu and H. Xiang, *Acta Pharm. Sin. B.*, 2021, **11**, 1300.
4. A. D. Buhimschi, H. A. Armstrong, M. Toure, S. Jaime-Figueroa, T. L. Chen, A. M. Lehman, J. A. Woyach, A. J. Johnson, J. C. Byrd and C. M. Crews, *Biochemistry*, 2018, **57**, 3564.
5. H. Singh and D. K. Agrawal, *Future Med. Chem.*, 2022, **14**, 1403.
6. J. Lu, Y. Qian, M. Altieri, H. Dong, J. Wang, K. Raina, J. Hines, J. D. Winkler, A. P. Crew, K. Coleman and C. M. Crews, *Chem. Biol.*, 2015, **22**, 755.
7. D. P. Bondeson, A. Mares, I. E. D. Smith, E. Ko, S. Campos, A. H. Miah, K. E. Mulholland, N. Routly, D. L. Buckley, J. L. Gustafson, N. Zinn, P. Grandi, S. Shimamura, G. Bergamini, M. Faelth-Savitski, M. Bantscheff, C. Cox, D. A. Gordon, R. R. Willard, J. J. Flanagan, L. N. Casillas, B. J. Votta, W. den Besten, K. Famm, L. Kruidenier, P. S. Carter, J. D. Harling, I. Churcher and C. M. Crews, *Nat. Chem. Biol.*, 2015, **11**, 611.
8. E. S. Fischer, K. Bohm, J. R. Lydeard, H. Yang, M. B. Stadler, S. Cavadini, J. Nagel, F. Serluca, V. Acker, G. M. Lingaraju, R. B. Tichkule, M. Schebesta, W. C. Forrester, M. Schirle, U. Hassiepen, J. Ottl, M. Hild, R. E. J. Beckwith, J. W. Harper, J. L. Jenkins and N. H. Thomä, *Nature*, 2014, **512**, 49.
9. G. Petzold, E. S. Fischer and N. H. Thomä, *Nature*, 2016, **532**, 127.
10. T. Ito, H. Ando, T. Suzuki, T. Ogura, K. Hotta, Y. Imamura, Y. Yamaguchi and H. Handa, *Science*, 2010, **327**, 1345.

11. C. Steinebach, H. Kehm, S. Lindner, L. P. Vu, S. Köpff, Á. L. Mármol, C. Weiler, K. G. Wagner, M. Reichenzeller, J. Krönke and M. Gütschow, *Chem. Commun.*, 2019, **55**, 1821.
12. J. Y. Hong, H. Jing, I. R. Price, J. Cao, J. J. Bai and H. Lin, *ACS Med. Chem. Lett.*, 2020, **11**, 2305.
13. E. F. Douglass, C. J. Miller, G. Sparer, H. Shapiro and D. A. Spiegel, *J. Am. Chem. Soc.*, 2013, **135**, 6092.
14. P. S. Portoghese, *J. Med. Chem.*, 1992, **35**, 1927.
15. Y. Sekido, S. Bader, F. Latif, J. R. Gnarra, A. F. Gazdar, W. M. Linehan, B. Zbar, M. I. Lerman and J. D. Minna, *Oncogene*, 1994, **9**, 1599.
16. Y. X. Zhu, C.-X. Shi, L. A. Bruins, X. Wang, D. L. Riggs, B. Porter, J. M. Ahmann, C. B. de Campos, E. Braggio, P. L. Bergsagel and A. K. Stewart, *Blood Cancer J.*, 2019, **9**, 19.
17. D. Senft, J. Qi and Z. A. Ronai, *Nat. Rev. Cancer*, 2018, **18**, 69.
18. H. Lebraud, D. J. Wright, C. N. Johnson and T. D. Heightman, *ACS Cent. Sci.*, 2016, **2**, 927.
19. K. Shanmugasundaram, P. Shao, H. Chen, B. Campos, S. F. McHardy, T. Luo and H. Rao, *J. Biol. Chem.*, 2019, **294**, 15172.
20. H. Lebraud, D. J. Wright, C. E. East, F. P. Holding, M. O'Reilly and T. D. Heightman, *Mol. BioSyst.*, 2016, **12**, 2867.
21. A. Bachmair, D. Finley and A. Varshavsky, *Science*, 1986, **234**, 179.
22. A. Varshavsky, *Annu. Rev. Biochem.*, 2017, **86**, 123.
23. T. Tasaki, S. M. Sriram, K. S. Park and Y. T. Kwon, *Annu. Rev. Biochem.*, 2012, **81**, 261.
24. R. J. Patch, L. L. Searle, A. J. Kim, D. De, X. Zhu, H. B. Askari, J. C. O'Neill, M. C. Abad, D. Rentzeperis, J. Liu, M. Kemmerer, L. Lin, J. Kasturi, J. G. Geisler, J. M. Lenhard, M. R. Player and M. D. Gaul, *J. Med. Chem.*, 2011, **54**, 788.
25. J. Y. Hong, H. Jing, I. R. Price, J. Cao, J. J. Bai and H. Lin, *ACS Med. Chem. Lett.*, 2020, **11**, 2305.
26. M. Schiedel, D. Herp, S. Hammelmann, S. Swyter, A. Lehotzky, D. Robaa, J. Oláh, J. Ovádi, W. Sippl and M. Jung, *J. Med. Chem.*, 2018, **61**, 482.
27. H. Liu, X. Ding, L. Liu, Q. Mi, Q. Zhao, Y. Shao, C. Ren, J. Chen, Y. Kong, X. Qiu, N. Elvassore, X. Yang, Q. Yin and B. Jiang, *Eur. J. Med. Chem.*, 2021, **223**, 113645.
28. F. Rossari, F. Minutolo and E. Orciuolo, *J. Hematol. Oncol.*, 2018, **11**, 84.
29. J. C. Byrd, R. R. Furman, S. E. Coutre, I. W. Flinn, J. A. Burger, K. A. Blum, B. Grant, J. P. Sharman, M. Coleman, W. G. Wierda, J. A. Jones, W. Zhao, N. A. Heerema, A. J. Johnson, J. Sukbuntherng, B. Y. Chang, F. Clow, E. Hedrick, J. J. Buggy, D. F. James and S. O'Brien, *N. Engl. J. Med.*, 2013, **369**, 32.
30. L. A. Honigberg, A. M. Smith, M. Sirisawad, E. Verner, D. Loury, B. Chang, S. Li, Z. Pan, D. H. Thamm, R. A. Miller and J. J. Buggy, *Proc. Natl. Acad. Sci. U. S. A.*, 2010, **107**, 13075.
31. J. A. Woyach, R. R. Furman, T.-M. Liu, H. G. Ozer, M. Zapatka, A. S. Ruppert, L. Xue, D. H.-H. Li, S. M. Steggerda, M. Versele, S. S. Dave, J. Zhang,

A. S. Yilmaz, S. M. Jaglowski, K. A. Blum, A. Lozanski, G. Lozanski, D. F. James, J. C. Barrientos, P. Lichter, S. Stilgenbauer, J. J. Buggy, B. Y. Chang, A. J. Johnson and J. C. Byrd, *N. Engl. J. Med.*, 2014, **370**, 2286.
32. J. A. Burger, D. A. Landau, A. Taylor-Weiner, I. Bozic, H. Zhang, K. Sarosiek, L. Wang, C. Stewart, J. Fan, J. Hoellenriegel, M. Sivina, A. M. Dubuc, C. Fraser, Y. Han, S. Li, K. J. Livak, L. Zou, Y. Wan, S. Konoplev, C. Sougnez, J. R. Brown, L. V. Abruzzo, S. L. Carter, M. J. Keating, M. S. Davids, W. G. Wierda, K. Cibulskis, T. Zenz, L. Werner, P. D. Cin, P. Kharchencko, D. Neuberg, H. Kantarjian, E. Lander, S. Gabriel, S. O'Brien, A. Letai, D. A. Weitz, M. A. Nowak, G. Getz and C. J. Wu, *Nat. Commun.*, 2016, 7, 11589.
33. S. Hati, M. Zallocchi, R. Hazlitt, Y. Li, S. Vijayakumar, J. Min, Z. Rankovic, S. Lovas and J. Zuo, *Eur. J. Med. Chem.*, 2021, **226**, 113849.
34. R. A. Hazlitt, T. Teitz, J. D. Bonga, J. Fang, S. Diao, L. Iconaru, L. Yang, A. N. Goktug, D. G. Currier, T. Chen, Z. Rankovic, J. Min and J. Zuo, *J. Med. Chem.*, 2018, **61**, 7700.
35. M. Hu, W. Zhou, Y. Wang, D. Yao, T. Ye, Y. Yao, B. Chen, G. Liu, X. Yang, W. Wang and Y. Xie, *Acta Pharm. Sin. B.*, 2020, **10**, 1943.
36. E. W. Yue, R. Sparks, P. Polam, D. Modi, B. Douty, B. Wayland, B. Glass, A. Takvorian, J. Glenn, W. Zhu, M. Bower, X. Liu, L. Leffet, Q. Wang, K. J. Bowman, M. J. Hansbury, M. Wei, Y. Li, R. Wynn, T. C. Burn, H. K. Koblish, J. S. Fridman, T. Emm, P. A. Scherle, B. Metcalf and A. P. Combs, *ACS Med. Chem. Lett.*, 2017, **8**, 486.
37. J. T. Madak, C. R. Cuthbertson, W. Chen, H. D. Showalter and N. Neamati, *Chem. – Eur. J.*, 2017, **23**, 13875.
38. Z. Sun, B. Deng, Z. Yang, R. Mai, J. Huang, Z. Ma, T. Chen and J. Chen, *Eur. J. Med. Chem.*, 2022, **239**, 114544.
39. W. Zheng, *Mini-Rev. Med. Chem.*, 2022, **22**, 2478.
40. B. Wang, S. Wu, J. Liu, K. Yang, H. Xie and W. Tang, *Eur. J. Med. Chem.*, 2019, **176**, 476.
41. M. Hanafi, X. Chen and N. Neamati, *J. Med. Chem.*, 2021, **64**, 1626.
42. K. A. Donovan, J. An, R. P. Nowak, J. C. Yuan, E. C. Fink, B. C. Berry, B. L. Ebert and E. S. Fischer, *eLife*, 2018, **7**, e38430.
43. S. Lindner, C. Steinebach, H. Kehm, M. Mangold, M. Gütschow and J. Krönke, *J. Visualized Exp.*, 2019, **147**, e59472.

Epilogue

In this book I have presented succinctly and comprehensively all the currently known concepts for the design of active site-directed inhibitors for all types of enzymatic reactions with carefully selected examples of inhibitors together with a delineation of their mode of working. Given the demonstrated power yet simplicity of implementing these concepts in quickly and cost-effectively finding effective active site-directed inhibitors for all types of enzymatic reactions involved in various life processes and therapeutic areas, this book, as the first of its kind in the field, will benefit its audience for quickly and efficiently obtaining effective active site-directed inhibitors for any of the enzymatic reactions under study. Besides being a reference book, this book could also be adopted as a textbook in a course for graduate students or upper-level undergraduate students.

Active Site-directed Enzyme Inhibitors: Design Concepts
By Weiping Zheng

Published by the Royal Society of Chemistry, www.rsc.org

Subject Index